Designing Religious
Research Studies

Designing Religious Research Studies

From Passion to Procedures

C. JEFF WOODS

WIPF & STOCK · Eugene, Oregon

DESIGNING RELIGIOUS RESEARCH STUDIES
From Passion to Procedures

Wipf & Stock
An Imprint of Wipf and Stock Publishers
199 W. 8th Ave., Suite 3
Eugene, OR 97401

www.wipfandstock.com

PAPERBACK ISBN 13: 978-1-4982-1892-4
HARDCOVER ISBN 13: 978-1-4982-1894-8

Manufactured in the U.S.A.   02/16/2016

Contents

Acknowledgments

I WISH TO THANK all of my former religious research students who provided the testing ground for this manuscript and improved the processes in innumerable ways. My deep gratitude is extended to Marilyn Tyson who provided the initial editing of the manuscript. My wife, Kandy, continues to provide tremendous support for my ministry and my children, Brandon and Kelsey, are a source of inspiration and joy. I am humbled and indebted to President Jay Rundell at the Methodist Theological School of Ohio and President Molly Marshall at Central Baptist Theological Seminary for opportunities to teach at their respective institutions. Finally, I express my thanks to Dean and former Doctor of Ministry Director Lisa Withrow and Doctor of Ministry Director Heather Entrekin, who encouraged me to pursue this project.

Introduction

The Journey Ahead

YOU ARE ABOUT TO embark upon a journey of religious research. At the beginning of this journey the vista is expansive and scenic. With each passing fork in the pathway and each subsequent decision, the journey narrows toward a manageable outcome. With this approach to research design, early decisions of topic and context demarcate later decisions of methodology and procedures. While the journey is in complete control of the researcher, sound research design requires that the researcher make key decisions regarding the area of study, data collection, and data analysis. This book is designed to make those decisions flow easily and naturally from the mind and passion of the researcher toward an outcome pleasing to the researcher and welcomed by the constituents. Consider this text your journey guide and the product the culmination of your passions as a researcher, for it is a grand journey that lies ahead.

A high percentage of religious studies students enter their research design class timorous of what the class might entail. Many bring their own personal horror stories in research design, statistics, and methodology. This approach to research design is intended to demythologize the research process. We conduct research every day when we decide which brand of cereal to buy, how to circumvent a detour, where to go on vacation, and who to take along. Research is ubiquitous. Shared patterns of research can be tailored toward outcomes that fit the researcher and satisfy the stakeholders.

This approach employs a funnel approach to research. As the researcher contemplates a topic of study at the top of the funnel, never again will so many choices be available to the researcher from this panoptic perspective. Passion should always drive one's choice of topic. Succeeding decisions about context, rising dilemmas, and supporting fields of study narrow the scope and limit the choices available to the researcher later in the process. By the time that the researcher names the research question, the very verbs contained in the question dictate appropriate methodologies and procedures to adequately answer the question. Whether or not the researcher will conduct interviews, design surveys, or facilitate focus groups will be completely ordained by decisions made at the top of the funnel.

This approach to research design has been used in multiple seminaries and institutions of higher education in both secular and religious environments. The process has been tried and tested with international students, traditional and nontraditional students, and multiple denominational contexts. The approach has been tested across theological and cultural boundaries.

This process not only produces a healthy outcome for students who choose to engage it, most students enjoy the journey. A seminary president recently introduced me as a person, "who teaches our students to love data." Personally, I never considered data anything but fun! One of the most common and certainly most pleasing comments that I hear from students of mine is, "I actually enjoyed this research class." For many years, students of mine have encouraged me to make this approach available to others in book form. I dedicate this text to all of my former students who have learned to love data.

Chapter 1

What is Research?

THE CONCEPT

WHAT IS RESEARCH? THE answer to that question probably has as many answers as readers of this book. In a moment, you will have an opportunity to create your personal response to this question, but before we dig into this question personally, let's explore the concept a bit further from a macro perspective. Religious research, like most research within the humanities, should take into account the personality, biases, and abilities of the researcher. Unfortunately, most approaches to research design ignore the distinctiveness of the researcher. This book will show you how to conduct sound, solid research that grows out of your passion as a person, and out of your context as a researcher. Conducting rigorous research is based upon using proven techniques. Sound research is not based, however, upon every researcher using an identical set of proven techniques. Unless you deal exclusively with petri dishes, your identity as a person and the topic you choose to study should strongly influence the way that your research is designed and conducted.

Design and Methodology

I have taught research design for over twenty years. Primarily, I teach research design to doctoral students within the religious context, but I have also taught research design to MBA students, educational students, and even to nursing students. By far, the most common flaw that I have encountered during my career is witnessing students being forced to fit their research topic into a particular set of research techniques. Effective research allows the research topic to drive the research design. That is the way it should be. Unfortunately, many programs try to fit all of their students into a single style or approach to research.

Research involves the collecting and analyzing of data for the benefit of self and others. The nature of what is being studied should always influence how it is studied.

Because of my background in research, many students facing their final research project have come to me seeking to dialogue about their project. I am always delighted to engage in such discussions. I am not always equally delighted by their revelations of the parameters that have been placed upon their research design. One student plainly told me, "Oh, we are not allowed to conduct any surveys or even come close to anything quantitative." Another student who was allowed to conduct quantitative research had already collected all of his data and asked me "to help him make sense of what he had already collected." Yet another said, "I am allowed to study anything I want, so long as I approach it from an ethnographic perspective." Another student disclosed to me that her freedom for research must be contained within a four-week intervention. Other students have been forced into a myopic "participant observer" approach regardless of topic, or have been told that focus groups are only for the marketing students, or have been instructed to conduct exactly six interviews as a part of their final project; not five, not seven, but exactly six. Apparently, in many research design classes, research methodology drives the research design and even the selection of one's topic. That order seems reversed to me.

Good research designs and methods are dictated by carefully chosen research topics. As the researcher makes decisions regarding the field of study, the context of the respondents, the nature of the anticipated discovery or development, and the texts and literature that ultimately shape the project further, the student should be carefully working their way toward a sound research design and set of research methods dictated by those earlier decisions.

The Funnel Approach

This book employs a funnel approach to research design. The top of the funnel is wide and representative of a student's beginning stages of selecting a research topic. During the beginning stages of selecting a research topic, students have every possible choice available to them. Never again will the funnel be this wide or contain this plethora of options. That is why it is important to ensure that the research topic emanates from the named passions of the student, only then to be followed by the definition of a problem, a formulated design, a set of procedures, and eventually a full-blown methodology. The specific research methodology occurs near the bottom of the funnel, clearly informed by each preceding decision. Research methods such as surveys, instruments, interviews, and focus groups are not baleful tasks to be feared, for they will flow naturally out of the chosen topic of the student. The reader who pursues this book in earnest will discover a wonderful journey where the research flows naturally from one's passion to the detailed procedures of the study.

Students pursuing this text in a classroom setting will gain an ability to recognize sound methodology for others, in addition to that for their own projects. At one point in the process, students will be able to accurately suggest whether a quantitative or qualitative approach is more fitting as a list of verbs from previous research projects are presented to them that call out for a further exploration of either the breadth or depth of information. You are invited to join the journey of designing a religious research study, from passion to procedures.

DIGGING IN

Word Exercise

If you have a few moments to perform this exercise, I encourage you to do so, or to perhaps come back to this exercise at a later time in your reading. In the "digging in" portion of this chapter, I invite you to perform a simple word association exercise. If you have any 3 x 5 cards or sticky notes handy, grab a stack of them and jot down every word or phrase that comes to mind when you think about the question, "What is research?" Write every word or phrase on a different card. If you do not have any 3 x 5 cards close, tear up a blank sheet of paper into some small pieces and use the pieces of paper like the 3 x 5 cards. During the next two to three minutes, write down as many different words or phrases that you can think of in response to the question, "What is research?"

Categorizing

In order to make more sense of what you have written down, begin to put the cards or pieces of paper into small piles according to their similarities. But, please use the specified process below to create the categories. It would be helpful if you could find a place with sufficient space to spread out the cards as you place them into the categories. Using sticky notes instead of cards to create groupings on the wall also works well for this exercise.

1. Pull out at random one of the cards with a word or phrase written on it.

2. Pull out the next card and ask yourself the question, "Is this word or phrase more alike or more different from the first word or phrase?"

3. If the second card is similar to the first card, place the second card beside the first one so that you can still read both of the words or phrases. (Do not worry about labeling the categories yet.)

4. Pull out a third card from the pile and proceed in a similar manner. If the third card is similar to the one of the first two cards or is similar to your newly formed grouping, place it next to that card or grouping and if it is more different than similar to any of the words or groupings created so far, put the card aside by itself.

5. Continue to read and place each card into an existing grouping or create a new grouping with that card. Proceed until you have placed all of the cards into groupings or have decided that one or more cards do not fit into any of the other groupings.

6. Now label each category. Begin with the grouping that contains the most cards and write down a word or phrase on a new card that portrays what the cards in this grouping have in common. It could be that one of the cards already in the grouping serves as an appropriate label for that category. Continue labeling each category until you have labeled every category that contains more than one card. ·

7. Next, take a look at the cards that seemed more different than similar to any of the other groupings. Now that you have labeled the other groupings, do any of these cards fit into the new groupings? If so, place the single cards into those groupings, leaving any singular cards that just do not fit well into any other place by themselves.

8. You now have created your own definition of research using a clustering technique.

Learnings

You probably discovered a few things about research as you developed your definition of it in the previous exercise. One of the first discoveries about research is that it usually involves a process. Seldom is research conducted in a single action. Most research involves a set of smaller processes within a larger framework or process. Definitions of research typically include some sort of planning process, some sort of data gathering process, a set of data analysis techniques, and a set of processes aimed at making sense of the entire study. Notice how the design naturally precedes the methods.

Good Research vs. Generic Research

Many former students who have completed this exercise distinguished between *generic* research and *good* research, making the leap to evaluative categories. A student may study any topic, but there are certain components that often make for better research. One of those components is that the researcher is passionate about the topic. Passion is also classically contagious. This passion of the researcher is often accompanied by the passions of other readers. Thus, a second quality of good research is that others are interested in its outcome. Good research is also timely. Not only are others interested in its outcome, but there is an urgency that accompanies good research. Many good research studies also delve into some sort of anomaly. Any insight into an enigma is welcomed, even if it does not fully explain the abnormality. We will discover other components along the way, but it is helpful to begin to distinguish good research from generic research from the outset.

Did you have any single card categories? Do not exclude them or ignore them. Learn to love *outliers*. Learn to pay attention to and embrace the surprises of your research, including those single data points that seem not to fit anywhere else and even go against the grain of your other collected data. Another quality of good research is that good research embraces outliers. Outliers are often the crucibles of great discoveries.

CLASSROOM EXERCISE

Each chapter of this book will contain a section on classroom exercises. In most instances, these exercises may be performed individually as well as in a group setting. Even if you choose not to work through the exercises, I encourage you to read through these sections in order to gain additional insight into the concepts presented.

Word Association

The 3 x 5 card exercise contained in the previous "Digging In" section works very well in a classroom setting. Use larger 4 x 6 sticky notes instead of 3 x 5 cards so that the students' responses can be posted to a wall. A wipe-off board works particularly well for this exercise. Just make sure that the notes will stick to the surface prior to conducting the exercise. I usually invite the students to work in groups of two or three depending upon the total size of the group. Invite the small groups to generate as many answers as they can to the question, "What is research?" Give the groups two to three minutes to complete their work.

At the end of the two to three minutes of generative work, encourage a representative from each group to bring their group's work to the front of the class or to the posting wall. Ask the representatives to form a line with their papers in hand. I usually maintain control of the posting process, but invite each group representative to announce one of their items and hand it to me and then proceed to the back of the line for their next naming and posting. I post the first note. For the second note, I ask if this new item is more similar or more different than the previous item and begin to form groupings according to the similarities and differences. Resist the temptation to label the groupings until all of the notes have been posted.

As an added incentive, I sometimes announce that this is a contest and that a prize will be awarded to the group that surfaces the greatest number of unique items. To avoid posting duplicate items on the board, if an item is challenged as being the same rather than merely similar, I pose the question to the entire class to determine whether a new item is a duplication of a previous item. Typically, the only items that are perceived by the class as a duplication of a previous item are those items that are indeed exactly alike or are another form of the word of a previously posted item. Remember that the goal is to group similar items together, so similarities are welcomed. For a prize, I usually give the group a container of healthy snacks, suitable for sharing with the entire class.

This sticky notes classroom exercise has the potential to greatly reduce group anxiety by demonstrating to the class that they already have a sound working definition of research in their heads. One particular Saturday morning, I picked up a box of donuts, not exactly a healthy snack, but my selection was limited at this particular convenience store. The doctor of ministry class that I was teaching appeared to be acutely apprehensive of taking a research class. Following the exercise, I attempted to reinterpret that reality by saying, "See you really don't even need me. You generated all of the content yourselves for a beautiful working definition of research." Upon hearing this statement, one of the students replied, "Oh yes, we do need you." As my ego began to rise a bit, the student finished the sentence, "Who would bring the donuts!" That was one of the best teaching moments I have ever had. The goal of all teaching is to create the space for the students to make the discoveries themselves. The classroom exercises contained in this book are designed to allow that to happen.

If time allows, a good group exercise is to fashion the product of the board groupings into a narrative definition of research. If time is limited, I form a narrative definition or ask one of the students to do so. Classes will often ask if they have left out anything important. I try to resist adding anything myself, but will some-times ask leading questions to encourage the group to round out the content if needed. The typical areas contained in most working definitions include:

- Preparatory work
- Research approaches
- Mindsets and attributes
- Collecting information
- Working with the gathered information
- Sense making
- Sharing the learnings

Analyzing Qualitative Data

The sticky note exercise is also a great introduction to the analysis of qualitative data. As students collect information from inter-views, focus groups, and observations, I encourage them to place each piece of information on a new card, coding the card with the information of when and how the information was gathered. The creation of categories from the gathered data can be done in a similar manner to the creation of groupings in the definition of re-search exercise. This realization of the potential application of this classroom exercise by the students often serves as an extremely powerful exhortation. So many students come into a research de-sign class worried about learning all kinds of data analysis proce-dures. When they realize that they have just quickly learned a very important qualitative data analysis procedure and had fun doing it, this exercise proves to be a terrific confidence builder for the group.

Anxiety Poll

Another exercise that I sometimes conduct at the beginning of a classroom experience is to ask each student to reveal a number between one and ten that represents their level of anxiety/comfort with the content of the research design course, where a one represents extreme anxiety and a ten represents extreme comfort. This exercise brings to light the diversity of anxiety levels contained in the group. It is good information for the group to have, and certainly helpful information for the instructor as well. I will show what else can be done with this data in a later chapter on analyzing quantitative data.

ADDITIONAL THOUGHTS AND RESOURCES

All research involves bias. There was a time when researchers believed that they could remove all biases and influences from their work, but many now believe that the mere observation of an object changes what we see. "When you change the way you look at things, you change the things you look at."[1] As a student works in the beginning stages of selecting a research topic, I recommend that the student explore their own personal biases associated with that topic and include those biases in their research design. A study that states the biases up front has more integrity than one that pretends that there are no biases. Stating bias is a natural part of allowing the research design to emerge out of the identity and context of the researcher. No matter how sound the techniques, there are no techniques that can remove all bias from the research methodology. It is better to state one's biases and allow the reader to keep those biases in mind when exploring the findings and implications of their work, than to assume that the research is value free.

---

1. The quote is widely attributed to Max Planck, a quantum physicist, 1858–1947.

Chapter 2

Passion

THE CONCEPT

Passion: The Top of the Funnel

THERE IS NO REASON why an academic research project should not connect with a student's passions. When a student is about to spend a hundred or more hours studying a specific topic, it seems that everyone would benefit from a connection between the student's energizers and the chosen topic of study, including the student, the advisor, the stakeholders of the topic, and especially the potential audience. One student's passion may be another student's anathema, but if a particular study strongly appeals to one student, there is a good chance that the same topic may appeal to others. Even in academic programs that still ask minions to research professors' corpuses of interest, students would do well to connect with professors whose interests align sturdily with their own.

This chapter will argue for the necessity of the researcher to understand their interests and passions as well as their skills and competencies as a basis for selecting a topic of study. Some of the approaches to apprehending one's identity and interests contained in this chapter may be new and others may be familiar. The approaches are intended to be illustrative rather than comprehensive.

The reader is encouraged to ponder these approaches and supplement them with their own interpretations of self.

As pointed out in the introduction, this text espouses a funnel approach to research design. Each decision that the researcher resolves narrows the funnel a bit further for the next subsequent decision. At the top of the funnel, at the level of passion, the entire domain of the world is vulnerable to the researcher. At no other point in the process will there be such a menagerie of choices available to the researcher. Thus, the researcher is encouraged to choose wisely, especially in the beginning stages of the research study.

What Keeps You Up at Night?

The first assignment that a student receives in my research design syllabus is to write a paper in response to the question, "What keeps you up at night?" Typically those areas of our life or ministry that keep us up at night are the sources of our passion. There is a reason that our body is willing to lose sleep over an issue; it is weighty to us. Getting in touch with the salient issues of our lives and ministries is vitally important to a researcher. That is because there may just be a strong connection between the weighty issues of our minds and the weighty issues of the organizations that we lead. Our strongest concerns are often closely connected to the leverage points of our ministry organization. We may sweat the small stuff, but we seldom lose sleep over it. The sources of our somnambulist tendencies are often the same areas that need to be addressed within the organization.

When detailing one's greatest concerns in working toward the completion of this assignment, there are a few other parameters that typically prove useful in this identification process. One such criterion is to turn the item into a question. Questions add fascination to the process. Additionally, students will eventually need to write a research question to frame their study, so forming the item that keeps you up at night into a question is good practice. Second, the question must be alive. By that, I mean that the question cannot already be answered by the researcher or someone else, or at least the solution must not be apparent to the writer. The best case studies are those that place the storyteller, and consequently the hearer, directly into the middle of the muddle. It is similar with forming an "alive question." Next, not only must the item have a future, it should also have a past. The item should have some history to it. Questions that keep us up at night seldom surface over breakfast that morning. I want to hear how the student has already struggled with the question and perhaps explored how others have grappled with the same quandary. Finally, it is important for the student to suggest possible approaches to the problem without tipping the scale toward proposed solutions. Speculating is good practice for building skills that will be needed later, for more specific research methods.

Some of the questions that I have received over the years have included:

- How does a person know that what they are asking for is what they truly need?

- How can grace be made free?

- What happens when we die?

- Will heaven be the same for everyone?

- How do people construct what justice means for them?

- What is it, that, when two people in the same situation have the same impulse to do something society would, by and large, deem that something that should not be done, enables one person to stop without doing whatever the something may be, and allows the other person to figuratively step over the line and do it?

- What is the source of entitlement on the part of small congregations that believe God owes them a full-time pastor?

- What are the implications of religious plurality upon congregations in America?

- What does it take for one generation to empower another generation in leadership?

- Is there really such a thing as church renewal?

- How can the church best spend the one hour a week it is given to shape the lives of children and families when it feels like they are operating like a pit crew for a weekly NASCAR race?

Leverage Points

Let's take a deeper look at why it is important to surface one's passions when selecting a research topic. Have you ever played the game Jenga? The game consists of several small blocks of wood that form a tower. The object of the game is to remove the pieces, block by block, without destroying the tower. As the game progresses, it becomes more difficult to remove a block without knocking down the other blocks surrounding it. That is because, as the game progresses, the remaining blocks represent the *leverage points* of the tower. Touching any of the remaining blocks influences every other block. It is equally important to concede that the blocks removed and discarded up to this point were having very little influence upon the tower's structure. One could remove, improve, do a 180, and return one of the early blocks to its original position without affecting the tower in any way.

Some research projects are like that. A student can spend hundreds of hours on a topic that ends with a, "So what?" rather than a "See what . . .?" at the end. Getting in touch with one's passions is a process that can draw us closer to the leverage points of the organization. Those areas of the organization that potentially keep the leader up at night, not only represent the passions of the leader, but they also typically represent the leverage points of the leader's organization as well.

Further Benefits of Identifying Passion

Another reason for discovering one's passions early in a research study is so that the researcher is not blindsided by them. If a researcher is not in touch with his or her passions, it can be tempting to try to sway the results, often unconsciously, toward those biases. Identifying and naming one's desired outcomes can prevent a researcher from skewing results.

There is yet another reason to select a study strongly aligned with one's passion and that is that God may have put that passion there. In Ephesians 5:16, readers of the text are encouraged to "Make the most of the time, because the days are evil." A more in-depth examination of this passage reveals words that might be better translated as, "Redeem the kairos." As many may recall, the Greek word *kairos* is not easily translated into the English language, but carries the connotation of time that is "pregnant or full of meaning." While the word is often translated as "time" it seems to carry a meaning that is antithetical to chronological time. In *kairos* time, more meaning is packed into a moment than seems metaphysically possible. A student embarking upon a rigorous academic study project will spend plenty of chronological time on the study, all the more reason to at least begin with a topic that carries a "fullness of meaning" for the student as well.

What Gets You Up in the Morning?

While identifying what keeps a person up at night serves as the first assignment in my research class, the first verbal response that I ask of my students once they get to the classroom is to respond to the question, "What gets you up in the morning?" Our passions surface not only from those items that bother us, but also from those items that bait us. Getting in touch with one's passions involves more than identifying one's concerns, it also involves reminiscing about one's call, remembering one's raison d'etre, recalling one's drive that caused her or him to say, "Yes" to God and "Yes" to God's ministry.

> Students who are able to successfully navigate
> the intersection between, "What keeps you up
> at night" and "What gets you up in the morn-
> ing," will usually find a solid foundation for a
> cohesive research project.

DIGGING IN

What Can You Not Shed?

This section is intended to help the reader identify their passions by identifying quintessential components of their identity. Perhaps the best place to begin exploring our interests and identity is with the givens of our lives. A part of our identity that we sometimes overlook surfaces from those portions of ourselves that we cannot shed, such as our ancestry, our birthplace, our birth order, and our personality preferences. Surfacing our roots is a great first step in uncovering our identity and this step alone often leads to a poten- tial area of study. Some of the best research topics carry a sense of returning to one's roots as a researcher. Those items that continue to drive our passions as adults are often those items that domi- nated our earlier years. What characterized your interests in your formative years? Take a moment to record those. From your earli- est recollections, have you always had a compassion for others, a propensity for leadership, an affinity for numbers, or a fondness for mysteries? These are all sources of identity that can give rise to exceptional research projects.

Over the years, I have occasionally served as a reviewer for sabbatical grant proposals. The best sabbatical grant proposals often represent a return to one's earlier days of both leisure and adventure. A reviewer can almost gain an emic view of a person's unfinished agenda through reading those proposals and the best proposals are often the ones that suggested a reconnection with one's adventurous adolescence. Sabbaticals are akin to research studies. Just as a return to one's earlier passions can drive a fruitful sabbatical, so can those same roots drive the selection of a research topic that carries great significance for the student. While the hours spent on a research study can be exhausting, the topic itself should carry a renewing energy to the researcher by allowing the researcher to gain access and to uncover truth for self and others through the investigation. Research surfaces truth and truth often emerges out of a return to our origins.

Personality Preferences

Another part of one's identity that cannot be shed is one's personality. Some of the strongest and most striking understandings of one's interests and passions are derived from sources that force us to make choices about who we are, or perhaps realize how the choices that we have made previously have affected who we are today. Understanding one's personality preferences is one of those areas. I believe that it behooves the researcher to at least consider one's personality preferences when exploring a research topic. As with all popularized constructs, the potential for abuse of a theory is often correlated with the popularization of that theory. While the Myers-Briggs Type Indicator (MBTI) is not without controversy these days, I believe that the core gifts contained within its theoretical base are helpful and serve as a good supplement for the identification of one's interests as a researcher. The MBTI was first developed by Katharine Cook Briggs and Isabel Briggs Myers in 1943 and is used widely in seminaries and religious institutions.[1] Below, I detail my understanding of the core gifts of the MBTI and suggest how those gifts might influence one's selection of a research topic. This particular analysis of Jungian theory continues to peep at our identity through the lens of those pieces of our personality that cannot be shed.

E-I: Extravert or Introvert. An extravert carries the gift of the breadth of things and thus, is often attracted to studies that are

1. One of the most prevalent distributers of the MBTI is CPP (www.cpp.com).

broad in scope. In contrast, an introvert carries the gift of depth and is more likely to be attracted to studies that are narrower in scope, and necessitate a more thorough exploration of the subject.

S-N: Sensing or Intuition. A sensing individual brings the gift of being in tune with the present moment and might be more attracted to studies that seek to understand the present reality and perhaps draw on a variety of senses in order to do so. An intuitive individual brings the gift of the future and thus might be more attracted to studies that deal more with the realm of vision and possibilities than with the realm of present situations and circumstances.

T-F: Thinking or Feeling. A thinking individual typically approaches a subject and makes decisions regarding that subject based upon logic and rationality and thus might find more resonance in projects that lend themselves to a logical flow and progression in the design. A feeling individual typically approaches a subject and makes decisions regarding that subject based upon what people value in a given situation and thus, might find more affinity in projects that explore deeply held values.

J-P: Judging or Perceiving. A judging person typically relies upon their ability to organize when approaching a new subject and thus might be happier with a research design that requires careful arrangement up front. A perceiving person typically relies upon their ability to remain flexible when approaching a new subject and thus might be better suited for a research design that requires a responsive approach to the subject or the ability to make adjustments along the way.

Archetypal Patterns

The MBTI is largely based upon concepts promoted by Carl Jung,[2] but the MBTI is not the only validated instrument based upon Jung's work. Another instrument based upon Jungian theory was developed much more recently by Carol Pearson and Hugh Marr

2. Jung, *Memories.*

and is based upon archetypal patterns. It is named the Pearson-Marr Archetype Indicator (PMAI). Unlike the MBTI that suggests lifelong patterns, archetypal patterns are thought to change as one grows and encounters new environments. Archetypes are like storylines that live themselves out in our personal and professional lives. Below is a brief description of the twelve named archetypal patterns in the PMAI. This particular resource is most helpful in aiding the researcher to discover which type of narrative is currently being played out by the role of the researcher.[3] Knowing what story you are living as well as what story you seek to live could have great influence upon the selection of your research topic.

1. Innocent—The innocent carries on the traditions of the organization and as such, might explore topics that seek to preserve the institution's best assets into the future.

2. Orphan—Because the orphan recognizes that disappointments are a part of life, he or she might pull together a group to embrace reality and make the best of a difficult situation.

3. Warrior—Projects that require overcoming great adversity would attract the warrior researcher.

4. Caregiver—Because a caregiver seeks out those in need of care as well as the vulnerable, a researcher living out this storyline would derive much satisfaction from projects aimed at partnering with and meeting the needs of others.

5. Seeker—Because the seeker believes that the organization's best hopes for the future lie in exploring other organizations, they would relish the opportunity to jump into the lives and culture of another place.

6. Lover—Many projects require bridging gaps, such as those between what the organization is now and what it might become in the future or those between an organization and its community. These would attract the lover, especially if they necessitated the building of new relationships.

3. One may purchase and self-score the PMAI through the Center for the Application of Psychological Type (www.capt.org).

7. Creator—A researcher living the narrative of a creator would be drawn to studies requiring new combinations of the existing physical and personnel assets of the organization.

8. Destroyer—A destroyer researcher would welcome the opportunity to release the old in order to make way for what is new, different, bold, and better.

9. Ruler—Because the ruler is drawn toward the creation of stable and strong organizations, he or she would be drawn to studies that draw heavily upon the tools of policies, procedures, and processes.

10. Magician/Catalyst—Projects that study shifts in organizational culture and consciousness or create the space for it, would draw a magician like a moth to light.

11. Sage—Researchers who view their role as the discovery of truth or the engagement of problem solving live out the narrative of the sage.

12. Jester—Jesters are given license to poke fun at even the most sacred of traditions or the largest of elephants in the room and there are studies that require that role to be played.

The MBTI and PMAI form two sides of a coin that can purchase needed space for the researcher to operate happily and productively out of their passion. In order to select a topic that connects with one's passion, one must know what those passions are. Too many researchers are not adequately in touch with their own passions and identity as persons and leaders prior to selecting a research topic. Passion is at the top of the research funnel and must be addressed before diving down into it any further.

CLASSROOM EXERCISE

Listening for the Passions of Others

I encourage students to share the results of their "What keeps you up at night?" paper in the classroom, but with a very specific set of instructions to the other students. While the student shares a summary of their paper, I encourage the other students to listen for possible passions and interests of the student who is presenting. Sometimes, I ask the presenter and audience to face away from one another, which seems to enhance the experience of listening for insights that are deeper than the content being presented. Over the years, I have discovered that students are very capable of performing this exercise for one another. They seem to get better at it with each subsequent student, which doesn't seem fair to the first presenter. To address this, I have sometimes allowed the group to return to the first few presenters, adding any new insights that may have surfaced.

During this exercise, a kind of synergy surfaces in which each student realizes that the full range of voices in the classroom are necessary in order to receive a comprehensive portrait of their own passions and interests. The synergy that surfaces is not unlike the kind of synergy that can surface from a well-designed focus group, a point that I try to remember to raise later in the course when we are discussing research procedures.

Our Main Thing

Here is another exercise that I use to surface one's passion and interests. It is a very simple exercise, and has served well to identify those activities that persons desire to do more of as leaders.

1. Jot down five to seven activities that consumed the majority of your time during the last week or month or unit of time.

2. Check the most important activity on that list.

3. List the barriers that seem to get in the way of doing that activity more often.

4. Try to surface one approach that you can take to reduce the barriers and engage in that activity more often.

ADDITIONAL THOUGHTS AND RESOURCES

The resources contained in this chapter are intended to be illustrative rather than comprehensive. I have attempted to include some of the newer resources that I have discovered for the documentation of identity and passion. Another resource that I have used over the years is *The Path* by Laurie Beth Jones,[4] which helps one craft a personal mission statement. The concept of fivefold gifts, beautifully articulated by J. R. Woodward in *Creating a Missional Culture*,[5] is another excellent exercise for surfacing one's passion and identity as a leader. A plethora of applicable exercises also are contained in William Kondrath's *God's Tapestry*.[6] For those looking for a more narrative approach to discovering one's inspirational path as a researcher, I would recommend Margaret Wheatley's, *Walk Out Walk On*.[7] Finally, for those searching for a comprehensive leadership assessment tool, I would recommend "The Leadership Circle."[8]

The hard work of identifying one's identity and passion not only gives rise to the topic of study, but these insights often find their way into the study itself. One of the places where one's passions should be inserted into the narrative of the study is in the realm of bias as touched upon in the previous chapter. All research involves bias. There was a time when researchers believed that they could conduct research studies void of their own preconceived thoughts and preferences. Few believe that these days. It is much better to surface one's biases up front and allow the reader to gauge whether or not such bias has been interjected into the processes of the study. But, recognizing one's bias and naming those biases are two very different activities. Biases can be extremely difficult to resurrect and name. Surfacing one's passions and interests greatly

---

4. Jones, *The Path*.
5. Woodward, *Creating a Missional Culture*.
6. Kondrath, *God's Tapestry*.
7. Wheatley, *Walk Out Walk On*.
8. Available at www.theleadershipcircle.com.

enhances the researcher's ability to name their own biases associated with the study.

A second place to insert one's discoveries related to passion, interests, and identity is into the rationale of the study. It is important to include why you have chosen to study this particular topic as a researcher, and more importantly, to make the case for why the subject needs addressing. Persons who spend adequate time bringing their passions to light have ample material for making a strong case for the rationale of the study.

Chapter 3

Context

THE CONCEPT

ON A PERSONAL LEVEL, defining one's context sets the scope of the project. From an audience perspective, defining the context of your constituency signals to the reader whether the results of this study might be applicable to their setting. Let's deal with the perspective of the audience first. The audience is interested in two primary questions regarding a study's context, namely what their own context may have in common with the researcher's context and how the researcher's context differs from their own context.

Thus, the researcher must achieve two aims
when describing the context of the study: detail
the commonality of the context as well as its
uniqueness.

Organizations Are Like People

In detailing one's context, it is helpful to be reminded of the fact that organizations are very much like people. Organizations have a past, a present, and a future—as well as a focus, a passion, an identity, a set of routines, and a set of idiosyncracies. Defining one's context is much like defining one's self. In fact, many of the theories and frameworks that apply to individuals also apply to organizations. For instance, the MBTI that we discussed in the last chapter has been applied to organizations, as has the PMAI. Organizations have personalities and storylines just like people. Although the theory that I am about to introduce in this chapter was not included in the last chapter on individual passion, it is a theory originally intended for individual identity and I would like to introduce it in this chapter as a framework for describing one's context of ministry. What we will discover from this theory is that the details of what allows people to fit in with other people are the same details that make one organization similar to another organization.

The theory comes from Ernst and Chrobot-Mason, who point out that much of our individual identity surfaces from two juxtaposed sources,[1] namely our desire to fit in with others and our desire to be unique. The categories that Ernst and Chrobot-Mason use to show how people seek to fit in with others work beautifully as a framework for describing one's organizational context. The authors declare that we describe ourselves in the five following categories:

1. Vertical categories using words such as rank, authority, and power

2. Horizontal categories using words such as departments, peers, and functions

3. Stakeholder categories using words such as networks, communities, and groups

---

1. Ernst and Chrobot-Mason, *Boundary Spanning Leadership*, 42.

4. Demographic categories using words such as race, gender, and generation

5. Geographic categories using words such as regions, languages, and global/local[2]

How Does Your Organization Compare to Others?

We learn from Ernst and Chrobot-Mason that responses to the question, "Who am I?" reveal the ways in which we are similar to other persons, namely vertical, horizontal, stakeholder, demographic, and geographic categories. As we will see in the following paragraphs these same categories disclose your organization's common features to other organizations. I encourage you to reflect upon your organization as these five personal categories are applied to institutions and congregations.

Consider first of all, your organization's vertical relationship with other organizations. If you are currently serving a congregation as a pastor or staff member, does your congregation have a hierarchical relationship with other congregations? Are you accountable to the Pope, a bishop, a district superintendent, a council, a presbytery, etc.? What is the nature of that relationship? If you are a hospice organization or other form of nonprofit, what does your hierarchy look like? As you reflect upon your organization's place in that system, what status do you have compared to others? Were you one of the founders or early members of the association or judicatory? How do others in the system view you? Are you an early or late adopter of denominational programs and initiatives? Answers to these types of questions can help define your vertical commonality.

What about other organizations horizontally? Are you part of a covenanted group, a federated group, or a network? Within that network, does your congregation or organization play well with others? Are you a leader among the other organizations? Whom do you consider to be your closest organization horizontally? Do

2. Ibid., 19.

you hold common worship services, collaborate on community projects, or engage in projects that cut across several other organizations horizontally? Responses to these questions define your horizontal boundaries and help other organizational leaders assess their commonality with you.

Defining your constituency is another aspect of naming your context. Religious organizations typically have three main types of constituents, namely their leaders, their servers, and those being served. What is your vision for each of these types of constituents? How many people fall into all three categories? How have your constituents, members, or congregants changed over the years? What would a timeline exercise of your organization reveal?[3] What have been the highs and lows of your constituency? What about your community constituents? Do you partner with people in the community or do you primarily meet their needs? Is your organization better at networking with similar or diverse organizations?

For the fourth category, we ask, who are you demographically as an organization? What percentage of people in the silent, boomer, buster, generation x, and millennial cohort groups are in your congregation? How does the average age of your congregation compare to other congregations and the general public? Who are you ethnically, racially, and theologically? What is your gender mix? Among which demographic groups does the power and energy reside? How does the demographic profile of your leadership group compare with your overall constituency as well as those whom you serve? How old is your organization and how does your tenure compare to others? Responses to these and other questions form your demographic identity.

In the fifth and final category, you are asked, "What can you disclose regarding your geography that will help others see what they have in common with you?" How many miles, yards, or feet away is the nearest similar organization? Have you always been in your present location? Are you urban, suburban, exurban, or rural? If urban or rural, exactly how urban and how rural are you? How

3. Ammerman et al., *Studying Congregations*, offers several forms for conducting a timeline exercise for your congregation or institution.

is your neighborhood shifting? What are the closest businesses to your location? What types of places do your stakeholders or congregants pass as they drive to your congregation or institution?

Answers to these questions give the reader a sense of who you are categorically and it is these categorical similarities that allow the audience to see what they have in common with you. But, that is only part of the answer to the question of who you are. Your audience is also interested in how you differ from other organizations. To answer that question, we must dig deeper into your identity and surface a different taxonomy.

DIGGING IN

Elements of Distinctiveness

We also derive a portion of our identity from words that make us unique. According to Ernst and Chrobot-Mason, our identities are derived not only from those categorical labels that we use to show how we fit in with others, but a portion of our identity also comes from a description of our exclusivity.[4] We cannot uncover our uniqueness by continuing to point to those factors that help us fit in with others. An organization's inimitability comes from sources of uniqueness rather than sources of collaboration.

An individual's uniqueness is derived from their attributes. When describing a person's distinctiveness we would tend to draw from adjectives that describe the person or we might name values that they hold dear, both of which are displayed through their behavior. It is similar with organizations. In order to discover the uniqueness of an organization, we must turn to the behaviors of that organization. While analyzing categorical differences assists others in discovering commonalities between their organization and your organization, it is through the analysis of the internal elements of an organization that its distinctiveness is divulged.

In order to dig in to the internal elements of your organization, I present a tool that I have used several times over the years

4. Ernst and Chrobot-Mason, *Boundary Spanning Leadership*, 7.

for organizational analysis. These elements were originally presented by David Clark in a classroom lecture and I have continued to use and adapt these categories over the years in various lectures and publications.[5] Prior to presenting a series of reflective questions as in the last section, the elements are described below.

1. Edificial elements—formally recognized components; the building blocks of the organization, such as worship services, small groups, committees, staff etc.

2. Functional elements—the ways that the organization accomplishes its stated mission such as worshiping, discipling, meeting needs, nurturing fellowship, serving, partnering, evangelizing, etc.

3. Procedural elements—the processes by which the organization gets things done that might include voting, consensus building, information sharing, assessment, evaluation, feasibility studies, needs assessment, staffing, budgeting, nominating, conflict mediation, etc.

4. Extra-organizational elements—how the organization interacts with other organizations denominationally, ecumenically, in the community, and internationally.

5. Idiographic—how the organization chooses to respond to changing cultures and environments.

Applying the Elements of Distinctiveness

The secret to the distinctiveness of your organization is probably contained in one of the elements listed above. Some organizations reveal their uniqueness through their edificial elements. Is there some feature that makes your worship or small groups different from other congregations? Does your congregation behave in a unique manner when worshiping or studying? During one of the deacon's meetings of the first congregation that I served as pastor,

5. Woods, *We've Never Done*, 47–48.

they suggested that the congregation rise whenever I entered the sanctuary. That would have made our Baptist congregation quite unique, but I declined the offer. One congregation may worship with pomp and circumstance and another with a down-home feel and style. A congregation in Chicago focuses its entire spring and fall educational programming on an annual theme, with in-house writing of the curriculum materials. What makes the building blocks of your organization unique from others?

Perhaps you consider your building blocks very similar to other organizations, but find distinct facets in your functional elements. Increasingly, congregations are setting themselves apart in the ways that they choose to serve others. One congregation posts their phone number as a hotline for people in the community to call when they feel lonely and staffs the number 24/7. Another congregation schedules rides to and from the hospital for everyone in the community, whether members or not. Paul Sparks, Tim Sorens, and Dwight Friesen detail several congregations that uniquely partner with community agencies to create better environments rather than merely meeting the needs of individuals in solo efforts.[6] A congregation in Milwaukee is known for its community partnerships, like many other congregations in the U.S., but their unique processes set them apart from others in the way that they choose to partner. Anyone desiring to launch a new ministry may garner two thirds of the support for that ministry from within the congregation, but the remaining one third must come from the community itself with no ties to the congregation.

For other organizations, it is their procedural elements that set them apart from others. Over the years, several of my students have revealed somewhat unique ways of getting things done in their congregations. One congregation operates solely out of a consensus model. Another requires that all persons elected to office must first identity their spiritual gifts, with no exceptions. Another requires all new members to take a conflict mediation class. A congregation in Topeka, Kansas recently adopted a new mission statement based upon interviews with persons in the community.

6. Sparks et al., *The New Parish*.

Consider the ways that decisions about people and money tran-spire in your organization. Are these behaviors a source of your distinctiveness?

One former student of mine described the exceptionality of his congregation in this way, "Every year in the last weekend of June, the 'Battle of Blountville' is re-enacted on the front lawn of the Blountville United Methodist Church. Guests on any given Sunday throughout the year may also catch a glimpse of the can-nonball house (so named because of the fact that it was hit by a Union cannonball during the battle), which is owned by Blount-ville United Methodist Church."[7] This congregation has a unique relationship with its community. Is there a fashion in which your organization interacts with your community that sets it apart as well? Is the local food pantry distributed through your congrega-tion? Do you host the Founders' Day picnic? Do you border on jingoism on the Fourth of July? In what ways do you traverse the community?

Finally, there is an "other" category for those unique elements of your organization that simply do not fit into any other place, such as the congregation whose pastor brings his toy poodle into the pulpit to accompany the sermon on the last Sunday of every month; or the congregation in Maine that holds two morning wor-ship services at 9:00 a.m. and 11:00 a.m. with completely different foci and messages from the pastor because the second worship service is actually the evening worship service but was moved to the daylight hours because the members could no longer drive at night. It is often the discrete elements of an organization that re-veal the most about its identity.

An element need not be entirely unique to help distinguish the organization from others, if the combination of elements adds color to the portrait of uniqueness. Mentioning your one-board system, your annual community needs assessment, your zero-based budgeting process, or your current pipe organ campaign feasibility study may not be entirely unique but may add to your

7. Paper presented by Charles Ensminger at the Methodist Theological School of Ohio, 2011.

portrait. The key in determining whether or not to include those items in your contextual analysis is if they contribute to the emerging landscape of the portrait. The description of your context should paint a cohesive portrait and subsequently give rise to the topic of your study out of that context. Remember that each decision in the funnel influences the next. By the time that you present your research question to the reader after portraying your rationale and context, the presentation of it should not jolt the reader into a "never saw that coming" reaction.

CLASSROOM EXERCISES

Listening for Commonalities and Distinctives

Presenting a summary of one's contextual analysis is a key part of research design. Over the years, I have discovered that the key to these classroom presentations is to ask the other students to listen attentively for the elements of commonality as well as the elements of distinctiveness. Even though this directive may have accompanied the assignment, somehow students are able to hear some distinctive in another presentation that they may not be able to portray through their own presentation. The practice of listening helps each student hone and reorganize their own summaries in ways that surface the distinctiveness of their contexts. Labeling two columns on a wipe-off or smart board as "commonality" and "distinctive" and giving each student an opportunity to either serve as the first generator of ideas in response to the presentation or serve as facilitator of classroom comments also helps to concretize the experience for the students.

Using Archetypal Patterns to Surface Culture

One of the most difficult aspects of one's context to grasp and portray to others is the culture of one's context. The best tool that I have discovered for grasping culture is the archetypal patterns tool described in the last chapter.[8] For use in a classroom exercise, I have created 8 ½" by 11" heavy stock cards with the names of each of the archetypal patterns on each of twelve cards. I simply ask the participants of the exercise to stand by the card that best represents the current culture of the organization and then the culture that they would aspire to as a stakeholder of that organization. This simple exercise has surfaced new understandings of culture in ways that other tools and exercises have been unable to do so.

ADDITIONAL THOUGHTS AND RESOURCES

Resources

I have introduced a skeletal outline above for defining the context of an organization. An abundance of additional resources are available to assist the researcher with defining their context. As highlighted in a previous footnote, the text by Ammerman, Carroll, Dudley, and McKinney contains a plethora of tools and exercises for digging into one's context. Many of my previous students have also praised Branson's work on appreciative inquiry[9] as well as Snow's work on asset mapping.[10] Both apply their tools to a congregational setting. David Roozen and James Nieman's work on *Church, Identity and Change*, offers a wonderfully advantageous historical perspective on denominations.[11] Kathleen Cahalan's resource, *Projects that Matter*, has also benefited numerous students of mine. Her text suggests that students analyze the conditions of their context when summarizing its distinctiveness to others.[12] Conditions may include unique needs, problems, issues, questions, and opportunities.

8. Pearson, Carol. *Organizational and Team Culture*.

9. Branson, *Memories*.

10. Snow, *The Power of Asset Mapping*.

11. Roozen and Nieman, *Church*.

12. Cahalan, *Projects That Matter*.

The penultimate goal in describing one's context is to create a "thick description" of that context. In previous eras of research design, researchers would seek to describe all of the ways in which their study might differ from other studies and suggest particular parameters that other organizations must hold in common in order to benefit from their findings. After all of the limitations had been exhausted, such a list of potential organizations was typically quite small. The newer method of describing context is through a thick description. If the description is thick enough and adequately details the commonalites and distinctives of the context, potential readers can conclude for themselves whether or not the study might apply to their setting.

The Scope of Your Study

As we have seen, portraying the commonalities and distinctives of your context can greatly benefit the reader by allowing the reader to discern which of the insights that have been made can be applied to a specific setting. Not only does the described context of your organization assist your readership, it also assists you as a researcher by defining the scope of your study. Whether you view your context as a single entity or as a collection of organizations will have a dramatic effect upon the scope of your study. If you are a district superintendent or a district manager of a group of YMCAs, you may have no choice but to view your context as a collection of entities, but in order to narrow your scope it might be appropriate to target a subset of those entities based upon your leadership interests. Recalling your passions can limit your scope. Is your leadership focus currently cascading toward struggling entities, growing entities, new entities, or turnaround entities? One of these foci may serve as a suitable pursuit for your project. The passion of your leadership should serve as a natural delimiter of the scope of your study.

Others who serve as a pastor of a local congregation, as a chaplain of a college or prison, or as the director of a single non-profit may also view their context as being comprised of a larger collection of congregations or organizations in their community. One former student of mine, working from a missional church perspective, views his context as a collection of small congregations in his local community. A former student who serves as a health care worker viewed his context as a collection of heath care agencies. A hospice worker identified the clergy and hospice workers in her community as her particular context. It is important to gauge one's context carefully prior to spending hours to describe that context. The decision will also set a fitting scope for your religious study.

Every decision that you make as a researcher further defines your study. When you uncovered your passion as a researcher, the world was at your fingertips. As you paint the portrait of your context you are further delimiting the parameters of your study. The good news is that you are scaling your study to a manageable size with each decision. The even better news is that by the time you get to the methodology portion of your study, which is often the most dreaded portion of the study, your methods and procedures will fall into place like one more natural piece to add to the puzzle that you are creating.

Chapter 4

Discovery and Development

THE CONCEPT

So far we have discussed the importance of religious studies rising out of one's passion and one's context. This chapter will demonstrate how to ensure that your project is intriguing to you and your audience. Most books that get written about organizations, including business books, organizational theory books, books for nonprofits, and books about congregations either show you something that has been discovered or show you how to improve something. Thus, nearly all research that gets funded and published about organizations centers on either development or discovery. This chapter will explore both of these concepts in detail.

Research designs that include one or both of
the elements of development and discovery
yield intriguing results regardless of what the
collected data reveals.

Rather than being dependent upon the data collected, intrigue can be built into the design of the study.

DIGGING DEEPER

Ever heard someone say that a date, a new friend, or an interviewee was "interesting?" Compare your feelings to that response with hearing that the person was "intriguing." There is a huge difference between the two. There also is a difference between an interesting research project and an intriguing research project. A project that arises out of your passions and context may be interesting to you, but still not interesting to others. To advance from interesting to the level of intrigue, the project must contain additional elements beyond passion and context. In the sections below, I will show you how to add an element of intrigue to your research design. Let's start with discovery.

Discovery

The difference between an interesting discovery and an intriguing discovery lies in the difference between solving an equation and solving a mystery. All discovery projects solve an equation. They find "x." They fill the blank in the sentence. They reveal something previously undisclosed. But the great studies solve mysteries. To solve a mystery, one must identify an anomaly.

It is startling to hear that ISIS has been recruiting terrorists from U.S. soil and British soil. It is perplexing to hear that three teenage girls from suburban neighborhoods have decided to join ISIS. Did you notice what shifted the news from startling to perplexing? Let's try another example. It is startling to hear that someone has lost their life in a drowning accident. It is perplexing to hear that the person was an adult and drowned in three feet of water. It is that second piece of information that creates an anomaly and turns the interesting project into an intriguing project.

Some of the most intriguing research studies take on the character of an anomaly. Anomalies begin with a statement and then add an interactive statement that creates a collocation with the first statement. It is the interaction between the two statements that piques the reader's curiosity and causes an eagerness to discover the findings. Here are some examples of some research questions based on enigmas from some former students of mine.

One student designed a research study to discover how a group of crisis workers are able to find meaning in their lives and cope with the stress of serving as a crisis worker. In the preliminary research, the student discovered a collocation in that the typical crisis worker had either served for over ten years or had served for less than three years. Studying how crisis workers find meaning would have been an interesting study, one that connected the student's passion and context. Studying what causes some crisis workers to burnout after a couple of years and what allows other crisis workers to serve for more than ten years shifted the study from interesting to intriguing.

Another former student planned to conduct a needs assessment in order to open a counseling clinic in a neighboring town. Preliminary research revealed that three other persons had attempted to do so and had been blocked by various agencies. With that piece of information, the study shifted from interesting to intriguing. We can see how an anomaly was built into this next research question from the very beginning, "What criteria do pastors consider when making difficult and complex decisions about leadership and what are the consequences of those decisions?" The researcher sought those factors that allowed persons in similar circumstances to behave differently.

The anomalies of discovery studies can take a variety of forms. The interacting statements can juxtapose two opposing attitudes or they may contrast theory and practice as in, "Since the Pareto Principle is widely known and documented, what prevents leaders from engaging the other eighty percent of the congregation?" Another common form of enigma comes in the form of a knowledge/practice conflict as in the following, "Even though First Congregation is known as a cohesive congregation, what contextual factors in First Congregation contributed to the disruptive behavior of person X that resulted in the tragedy on X date?" The research question might also carry within it two seemingly contradictory practices as in, "Why do the congregational search committees in X district consistently present a distorted perspective of their congregation to potential pastoral candidates?" or in the question, "Why have the congregations in X judicatory spent so much time crafting mission statements and so little time implementing them?" A great number of research studies have as their primary aim the goal of discovering something unknown. To the extent that the discovery can reveal findings related to two contrasting factors, the study can advance from merely interesting to extremely intriguing.

Research Steps in Discovery Projects

I have outlined a procedure below for pursuing discovery projects that are based on exploring an anomaly. No research study will follow these precise steps, nor the steps outlined in the next section; rather the steps are intended to portray an overall approach to this type of study.

1. State the anomaly in the form of two contrasting statements.

2. Formulate the statements into a single research question.

3. Reshape the question based upon your theological and contextual analysis.

4. Speculate on the causes of the anomaly and why it continues.

5. Make a list of the most promising speculations.

6. Conduct a literature review to see which speculations have already been addressed by studies in similar contexts.

7. Design a methodology to explore each of the remaining speculations.

8. Collect your findings.

9. Share your findings.

10. Create a set of new speculations for the next researcher.

Development

This chapter is about designing research studies that are intriguing to the reader rather than merely interesting to the researcher. Earlier, we discussed the fact that most research studies either seek to discover something or develop something and we covered an approach to research that can move a discovery project from interesting to intriguing. Now we cover an approach that can similarly move a development project from interesting to intriguing.

The difference between an interesting development design and an intriguing development design lies in the difference between leveraging an organizational development aspect that lies above the soil, versus leveraging an organizational development aspect that lies beneath the soil. In an earlier work of mine on organizational development,[1] I contrasted twelve aspects of organizational development by compartmentalizing them into two camps, namely those items that are visible above the soil versus those items that are mysteriously hidden beneath the soil. There is that word "mystery" again.

Developing visible aspects of an organization is interesting. Developing hidden aspects of an organization is intriguing. Certain aspects of organizational life are much more difficult to address because they are more hidden, but when attended to these less visible aspects can add great capacity to an organization. For instance, it is ineffective to realize a new vision as an organization without realizing a new identity. Vision is that above the soil aspect of an organization that becomes visible to all of its constituents. But, the hard work of envisioning lies in developing a new identity. A new vision will never be realized without the tough, beneath the soil toil of naming conflicting values and developing new habits and practices. New identities require hard work, wrestling with new ideas, exploring new landscapes, arguing over losses, and negotiating competing values. Identity is a beneath the soil organizational dimension, while vision is an above the line organizational dimension.

1. Woods, *On the Move.*

The table below shows a two-column representation of the twelve dimensions of organizational development. The first column represents the beneath the soil dimensions, while the second column represents the above the soil dimensions. Each row symbolizes a different aspect of organizational life.

Dimensions of Organizational Life		
Organizational Aspect	Beneath the Soil Dimension	Above the Soil Dimension
Culture	Identity	Vision
Purpose	Mission	Results
People	Potential	Performance
Assets	Capacity	Allocation
Change	Passion	Maturity
Conflict	Perspective	Momentum

Most development projects will narrow their focus toward one or more of the aspects of organizational life in the left column such as culture, purpose, people, assets, change, or conflict. Each of these organizational aspects carries with it a below the soil dimension as well as an above the soil dimension. All research studies that delve into development yield interesting results, but, over the years, I have observed that projects that leverage "beneath the soil" dimensions are far more intriguing than those that leverage "above the soil" dimensions.

We have already seen how a project that concentrates on the culture of an organization will yield far more intriguing results if it focuses on the identity of the organization rather than merely the vision of the organization. This same phenomenon holds true for each aspect of organizational life. A study that converges on the need for change will yield more intriguing results if it seeks to understand the passion of the organization to change rather than one which merely adopts a set of procedures aimed at maturing the organization. Likewise, development projects that address organizational conflict will be more intriguing if they work beneath the soil in the realm of perspectives rather than merely above the soil in the dominion of momentum.

Below, I have listed a sample religious research question from former students of mine for each of the beneath the soil dimensions of organizational development.

1. Identity: "What impact does utilizing a liberation theological motif experientially, pedagogically, and reflectively have upon identity, vocation, and intentional community involvement in X college?"

2. Mission: "Can overseas mission act as a changing agent in building and shaping future leaders for X congregation?"

3. Potential: "How can mentoring be utilized to foster effective lay leadership in order to provide leadership in areas not currently addressed by X diocese?"

4. Capacity: "Can asset mapping shift the focus of X congregation from survival to engagement with the community?"

5. Passion: "How will the realization and understanding of its founding story shape the desire for change at X congregation?"

6. Perspective: "Does making a connection between curriculum and sacramentality help students experience God in the classroom at X school?"

Research Steps in Discovery Projects

I have outlined a procedure below for pursuing studies that are based on exploring beneath the soil dimensions of organizational development.

1. Identify the problem or opportunity you are seeking to improve or enhance within the organization.

2. Identify the organizational aspect most closely associated with that topic and the organizational dimensions that address that aspect.

3. Answer the question, "How will you know that development has occurred?"

4. Formulate a research question that includes the desired outcomes.

5. Re-shape the question based upon your theological and contextual analysis.

6. Find the leverage points of your organization, those areas that are most likely to influence your desired outcomes.

7. Conduct a literature review to see which of those leverage points have already been addressed by studies in similar contexts.

8. Design a methodology to develop your leverage points.

9. Form a team to help you engage the study and collect results.

10. Share your findings.

CLASSROOM EXERCISES

Discovery

It is relatively easy to demonstrate the difference between interest and intrigue in a classroom setting, especially for discovery projects, by drawing upon a local news story. As preparatory work, I typically do a little bit of research related to an unsolved case in the local news. I try to pick a case that is very recent and involves a very visible portfolio such as theft or murder, but one that also does not push any political hot buttons for the students, lest the class become distracted by issues other than learning about discovery projects. I begin by presenting a fact or two about the case, ones that are completely descriptive and consistent. I then begin to add facts that seem to contradict the original facts leading to an anomaly. Asking the students where the case became intriguing will usually produce a direct connection between contradictory statements and intrigue.

The local news story can also be taken a step further. Asking students how they might go about solving the case usually produces an outline very similar to the, "Research Steps in Discovery

Projects" detailed above. Students naturally speculate on what might have happened or what might be continuing to happen in the case. Speculations lead to actions plans and soon the students have produced their own outline for pursuing discovery research studies.

Development

In teaching students about development projects in the classroom, I print each of the twelve dimensions of organizational development listed above on a heavy piece of card stock, 8 ½ by 11 inches, and lay the cards out in a fashion that depicts the nature of above the soil and beneath the soil aspects of organizational development. I have used various formats for the depiction of the dividing line between above and below the line items, including a simple horizontal line, opposite sides of the room, and a bell curve to depict the life cycle of an organization. As students explore real life examples of areas that need improvement within their context, the cards become a backdrop for the conversation and the formation of development projects.

ADDITIONAL THOUGHTS AND RESOURCES

Lag and Lead Indicators

One way of describing development projects is through the use of the concepts known as "lag" and "lead." Lag indicators, sometimes called "dashboard indicators," are representative of the desired outcomes of a project. The dashboard of your vehicle contains several instruments including the speedometer, the gas gauge, and the oil light which show the overall condition of your vehicle. Some of these, like the speedometer, can change very quickly. Others, like the gas gauge, normally change more slowly. Indicators reveal the condition or health of your organization and should be measurable or able to be monitored in some way.

For instance, lag indicators for worship might include the overall health or condition of the worship service. Measuring worship attendance can be done very simply by counting the number of people who come to the worship services each week. Worship satisfaction can be measured by distributing a brief five question worship satisfaction survey to a random group of worshipers each month and charting the results. Lag indicators can be identified by answering the question, "How will we know if we have made a difference in our organizational development."

Lag indicators derive their name from that fact that they will normally lag behind other activity. It can take a long time to influence a lag indicator. In fact, most lag indicators cannot be influenced directly. For example, the only way to influence worship attendance directly is to drag people off the street and into the sanctuary just prior to worship, a strategy that might have detrimental unintended consequences. Lag indicators lag behind because they are often the result of several other factors, known as lead indicators. It is unhealthy to be too concerned about lag indicators. It is also not biblical. Scriptures such as Psalm 127 remind us not to place too much concern in lag indicators such as overall growth, giving, and conditions. Rather, we are encouraged to be faithful in the little things and leave the results up to God.

Lead indicators are sometimes called "mission drivers," because they are expected to drive the other dashboard or lag indicators. Lead indicators influence lag indicators. Continuing with the worship theme, a team focused on this aspect of development might identify three or four lead indicators that they believe will lead to an enhanced worship experience. A sample lead indicator for the lag indicator of worship satisfaction might be to conduct an exercise with a representative group from the congregation to identify a large grouping of songs that the group believes resonate with the entire congregation. Another lead indicator of this nature might be to involve a wider variety of worshipers in the leading of worship or to launch a worship team to work on weekly themes and desired outcomes for worship. Because accountability is the key to development projects, each lead indicator should be assigned to a particular person willing to be responsible for that lead indicator. The salient point is that one person should assume accountability for each lead indicator, because this is where the changes will occur.

No one, however, should be held accountable for the lag indicators. While each specific lead indicator or mission driver should be assigned to a particular person who will be held accountable for that activity, those persons should not be held accountable for how much the lead activity influences the lag indicator. It is hoped that the set of lead indicators will positively influence each corresponding lag indicator, but those results are left up to God. Remember, "Paul planted, Apollos watered, but God gave the growth" (1 Cor 3:6).

Balancing Discovery and Development

There seems to be a trend toward emphasizing development projects over discovery projects in Doctor of Ministry programs. I clearly understand the current emphasis upon deliverable outcomes and measurable change. But, restricting students entirely to development projects, in my mind, restricts the options of the student researcher far too soon in the research design process. The reality is that many projects come bundled, calling for an element of both discovery and development within the same project. It is important, however, to understand the nature of each approach prior to combining them or seeking to emphasize one over the other.

I have been able to strike a compromise for students in one particular program that sought to demand that students limit their projects to a development approach. The compromise was achieved by adding a teaching seminar at the end of the discovery process so that the student researcher can demonstrate how their findings might be applied to development issues in similar contexts. This seems like a reasonable compromise for programs that have promised deliverable outcomes to parts of their constituency such as a group of trustees, an accrediting board, or even the students themselves.

Given the realization that development and discovery are both ubiquitous in today's world, it seems unreasonable to eliminate one of them from any form of research. Just read any blog or pick up any newspaper to confirm that development and discovery weave their way into our lives on a daily basis. Further reflection upon these blogs and articles, however, teaches us that not all articles are created equal. Those that add an element of mystery to their approach are those that intrigue us most as readers. Development and discovery are a part of life. When studying life and its religious dimensions, it is possible to address the mysteries it contains, rather than merely the interesting aspects of it.

Chapter 5

Theological and Theoretical Background

THE CONCEPT

IN NOVEMBER OF 2008 I had the opportunity to visit with Etienne Wenger, author and cofounder of the Communities of Practice movement, at his home in California. "Communities of practice are groups of people who share a concern, a set of problems, or a passion about a topic, and who deepen their knowledge and expertise in this area by interacting on an ongoing basis."[1] Communities of practice is a model for encouraging practitioners to engage with one another and hone their skills in a particular area. An effective community of practice must have a domain, a community, and a practice. As we explored this topic, I noticed that Etienne collected album covers of rock bands that were displayed around his house. I posed a casual question to Etienne as we were trying to clutch the concept at hand, "Can a rock band be a community of practice?"

"No," Etienne replied and after a pause added, "But a group of bass players could be."

"What about a group of drummers?"

"Certainly."

1. Wenger et al., *Communities of Practice*, 4.

"What about a group of roadies?"

"That would depend."

"Upon what?"

"Upon their motive for engagement."

As this conversation ensued, I began to grasp the importance of defining a domain in pursuing one's passion as a practitioner. A community of practice is not about teamwork or group dynamics; it is about zeroing in on a specific practice that the group wants to refine. The same level of importance and focus can be transferred from the practitioner to the researcher.

The literature review that the researcher must conduct is not just about finding relevant resources; it is about zeroing in on a particular field of literature as a lens through which to view one's study.

In this chapter, we will explore the notion of defining one's domain for a religious research study and allowing that domain to be shaped by theoretical and theological resources.

Consider how readily a focus comes to mind for researchers pursuing other fields of study. The sociologist focuses on groups. The psychologist studies the mind. A musician has an instrument, and an archeologist has artifacts. A fashion designer has clothing to consider, a facilitator the process, an attorney the law, and a physician the body. On the other hand, what is the domain of a minister? Is it God? Is it life? Is it key events in one's life? Defining one's domain is a prominent element of the research funnel for all researchers, yet more difficult for religious researchers to name given the nature of ministry.

Those who work in the field of ministry have been called the last general practitioners on earth. It seems that clergy are expected to learn from nearly everyone else, except one another. Many clergy are expected to stay current with the latest developments and regulations in finance, management, psychology, counseling, history, worship, spirituality, leadership, community organizing, and many other domains. Because of the superfluity of intersections with other disciplines, the identification of one's domain is more important for the religious researcher than for any other type of researcher. Below, we will identify some clues for identifying one's domain in the field of religious studies.

DIGGING IN

Theoretical Research

It has long been debated whether or not religion is a multidisciplinary subject, made up of key components from other fields, or whether religion is its own field of study. Both of the arguments for either a core of religious ideas or a core of multidisciplinary fields breaks down quickly as one peruses the listings of required courses in preparatory institutions for clergy including seminaries, Bible colleges, and less formal pastoral training tracks. What is included and excluded from these lists defies consistency. Do we include Greek and Hebrew? Some curriculums focus entirely on biblical studies, while others emphasize management and leadership. Church history and biblical criticism are requirements in most places, yet seminary graduates often say that they need more emphasis upon the practical areas.

It would seem that the choosing of one's topic of study for a religious studies pursuit would help narrow the domain, but that is not always the case. Imagine that someone has identified the topic of organizational change as a topic of focus having worked through the research funnel thus far in the process. Upon doing the literature review for that topic, does that person pursue group dynamics? Perhaps worship or community networking should become the focus. What about psychology or sociology? Church history may also shed light on this subject. As one begins to define a topic for their religious studies pursuit, that topic may intersect with a vast variety of fields. Some help is clearly needed in this arena.

Perhaps theology comes the closest to serving as the core of religious studies, but theology appears to be more of a cornerstone of the foundation, rather than the entire foundation itself. To limit the literature review for religious studies to the realm of theology, or even to the realm of biblical studies, would be to view religious studies from one of many lenses that have been foundational to previously published and groundbreaking studies. For now, let's agree that theology must be included, but will probably not serve as the entire domain for most religious research studies. What else should be included in one's literature appraisal along with theology?

Drawing from the classic philosopher of religion, P. H. Hirst, Chryssides and Geaves suggest the following, "Another option is to regard the study of religion as a field constituted by various interrelated academic disciplines. The study of any religion is likely to involve at least four of Hirst's fields: morality, since each religion has an ethical code to teach; the arts, since religions have their literature, music, painting, and sculpture; history, since religions typically tell the story of humanity's plight and struggle through time towards liberation and salvation; and the human sciences, since religions involve people as individuals, communities, and sociopolitical organizations."[2] Rather than suggesting that all religious studies must draw from these four fields, I find it helpful to lift these up as optional domains from which to draw in one's pursuit of religious research.

Centering upon one of these four fields: morality, the arts, history, or the human sciences, can help define the domain of your religious study. Allow me to illustrate by demonstrating how a single topic may be viewed through each of these four lenses. I recently had a student who named the topic of plagiarism as her field of religious study. Certainly, the field of morality might serve as an appropriate domain for this study. In her literature review, she could study current criteria for ethical decision making and suggest how this topic intersects with other topics in the morality

2. Chryssides and Reaves, *Study of Religion*, 42.

landscape. The student might also cite trends and considerations that have recently been identified in the field of morality.

Morality is not the only domain that could be tapped for this study. The student might choose the domain of the arts for this religious study. More specifically, the researcher might focus upon worship and how recent demands and trends in worship have influenced this topic. I could imagine both sides of this issue; that perhaps the increased access to sermon material has created a platform for less adherence to the rules of plagiarism related to preaching resources. I could also imagine that same notion of increased access to resources increasing adherence to the rules given the possibility of plagiarizers being discovered. The sharpening of this lens would require an analysis of the arts material.

History could also serve as a suitable domain for this topic. The student might study the history of plagiarism as different media developed through the ages. This lens might lead to developing a typology of how plagiarism has been dealt with throughout history. Finally, the field of human sciences may also serve as an appropriate domain for plagiarism. How have recent sociological or societal trends impacted this topic? What might be going on psychologically in the mind of the minister, or sociologically in the congregation, as a pastor makes the decision whether or not to cite one's resources?

These four domains identified by Chryssides and Reaves provide practical choices for the identification of the study's domain. As we have seen, once one of these four areas is identified, the study can then delve more deeply into that area, specializing further within a particular domain. While it may be beneficial to draw from more than one domain, it is advantageous to see one's topic through a primary lens. Selecting that lens is part of the funneling process of conducting a religious research study.

Contextual Theology

Earlier, I suggested that theology must be given consideration when conducting religious research studies. We turn our attention now toward the format for that inclusion. Because context is so important to research in general, and particularly vital for religious studies, one of the key components in the inclusion of theology into one's religious study is to identify one's perspective of how God and context relate to one another. Considering how God is at work in one's context is an excellent place to start when addressing the aspect of theology within the research funneling process. Stephen Bevans in his book, *Models of Contextual Theology*,[3] identifies six different perspectives for understanding how context and theology interact. In using this typology with my research design classes, I have found that most students can readily identify with one, or perhaps a combination of the models set forth in this typology. Viewing theology through one's context has proven to be an effective method of allowing theology to shape one's religious research study. I will briefly describe each model below and give a sample research question associated with the model from a former student.

The first model identified by Bevans is the translation model. The researcher holding this perspective views the culture as basically good, but in need of the Gospel and believes that it is her duty to translate the Gospel into the existing culture. A biblical example of this model might be Paul's address to Athens in Acts 17. Through this lens, the better that the researcher understands the culture, the more effectively she is able to insert the Gospel into it. The positive side of this model is that it takes the Christian message seriously. The negative side is that it can harbor naïve notions about the culture. A sample research question from this perspective might be, "What doctrines and practices of the Friends (Quaker) tradition does X Friends Meeting wish to claim for its identity and pass on to future generations?"

The anthropological model is the second model lifted up by Bevans. Ethnographic studies are a common form of this approach. The presupposition of this model is that it is in human nature that we find God's revelation.[4] Rather than inserting the

3. Bevans, *Models of Contextual Theology*.
4. Ibid., 56.

Gospel into the culture as in the translation model, this researcher attempts to pull the Gospel out of the culture through the research methodology. The story of the Syrophoenecian woman portrays this model. Studies that seek to surface identity may also hold this perspective. A sample research question might be, "How can Furnace Street Mission help crisis workers find meaning in order to better cope with the stress of their workplace?"

The praxis model also views the culture as good, but slightly distorted or in need of liberation. Research methodologies that incorporate an action learning approach might take on this perspective of theology.[5] The presupposition of this model is that it is "faith seeking intelligent action,"[6] and the proponents of it might cite James 1:22 as one of their foundational texts, "But be doers of the word, and not merely hearers who deceive themselves." A sample research question from this perspective was crafted by a former student who penned his question in this way, "Can an emphasis on eating, playing, listening, praying, serving, and suffering together build Christ-centered community in a suburban congregation?"

Bevans' synthetic model is perhaps the most difficult to grasp as the title might imply merely a combination of other models, but it is more than that. This model derives its name from viewing culture itself as synthetic as it, "reaches out to insights from other people's contexts—their experiences, their cultures, and their ways of thinking . . . it makes creative use of whatever is at hand."[7] This model utilizes both the uniqueness and the commonalities of its subjects. A sample research question might be, "How does an intentional interim minister (IIM) marshal an environment conducive to addressing adaptive challenges?"[8]

The transcendental model is the first model that seeks a theological transformation of the culture itself. The culture is still seen as good, but in need of transformation. In this model, the culture

---

5. Marquardt, *Action Learning*, would be one example.

6. Bevans, *Models of Conceptual Theology*, 73.

7. Ibid., 88.

8. "Adaptive challenges" is a term coined by Ronald Heifitz in his numerous works on organizational change.

is in need of new wineskins (Mark 1:22), and theology is viewed as both activity as well as cognitive reflection. The starting point for this model is one's own religious experience and one's own life experiences. A former student's research question portrays this theological perspective, "Does making connections between sacramentality and the curriculum help students experience God in the classroom?"

The countercultural model is the final model presented by Bevans. In this model, the researcher views the culture as resistant. This model is the most prophetic and confessional of all of the models and views culture as radically ambiguous and insufficient. Proponents of this model may seek to interpret, critique, and challenge their context. We can hear the student's embracement of the countercultural model in the following research question, "Will a shift in the church's focus from the Great Commission to the Great Commandment increase their passion for, and willingness to participate in, missional outreach activities?"

As we have seen, studying the relationship between culture and context is a great launching point for inclusion of one's theological view in the religious study. Each chosen contextual lens carries with it a history of theological insights and narratives that may be further drawn from as one frames the theological aspect of their study. As one seeks resources for drilling further down into the relationship between context and theology, Bevan's text is an excellent resource for identifying the relevant theological acumens and authors.

As I mentioned, context is not the only dimension in which theology reveals itself in a research study. In addition to the relationship between theology and context, the researcher's approach to the study, the methodology, and the desired outcomes of the study are also key points of inclusion for one's theological tenets. Virtually any aspect of the study can be shaped by a biblical and theological narrative. All religious researchers already possess a vision of how their context and theology interact, but some effort in surfacing and naming that perspective may also be involved. Although it may not be the entire foundation, theology is a part of every religious study and the researcher will benefit from an analysis of the interface between the culture and the Gospel.

CLASSROOM EXERCISES

Research Abstracts

A classroom exercise that I have practiced with good results over the years is to enter the classroom with several copies of the morning newspaper and to begin to distribute various sections of the newspaper to students according to their desires for reading the sports section, the business section, the arts section, etc. Once every student has a section in front of them, I ask them to find one article of interest and within that article identify the following three items from the article:

1. The main point of the article

2. What research was done to reach that conclusion

3. Suggested next steps in research

I give the students five minutes to complete this assignment and ask each person to report out in thirty seconds or less. Then I distribute copies of a recent research journal, the selection of which depends upon the interests of the class. Whichever journal is chosen for distribution, it is important to select a research journal that includes an abstract at the beginning of each article. With a copy of a research journal in front of them, I then ask the students to repeat the assignment identifying the main point, the research conducted, and next steps from a chosen article in their copy of the distributed journal. I then spawn discussion about the process of their work rather than the content generated regarding which assignment was easier: perusing the morning newspaper or perusing a research journal. Most conclude that is easier to peruse a research journal, especially after they discover that an abstract contains all of the information requested.

To continue the exercise, I often walk the class through an exercise to calculate how many articles they think that they could review in a three hour period (half a day), in order for the class to reach a collective conclusion about how long it might take to do their theoretical and theological literature review. Most are surprised by the brevity of time that it might require to get a handle on this aspect of the project. Bringing in an information librarian associated with the institution is also very beneficial at this stage of the process.

Text Shaping

Another popular exercise that I have conducted over the years involves identifying biblical texts that might shape the students' research topics. Depending upon class size, I form small groups of three or four students each and ask them to take turns describing their current topic and then receiving suggested texts from the other persons in the group that might shape that topic biblically. I have witnessed some fascinating results from this study as students suggest texts and theologians to a fellow student that they may never have surfaced themselves. On more than one occasion this exercise has resulted in the identification of the primary text for the student as they further develop their study and rationale. I have never had a student float a topic for which others could not pinpoint texts that might shape it. I suppose if that ever happens I might be tempted to say, "Perhaps that topic does not belong within the realm of religious studies."

ADDITIONAL THOUGHTS AND RESOURCES

While conducting a formal literature review at the beginning stages of one's religious study is the most common time in which to access theoretical and theological resources, it should not be the only time that this process is done. Interactive theories can surface at the beginning, middle, and final stages of one's study.

As a student delves into their methodology, occasionally questions arise that prompt a supplemental literature review. As a theory begins to surface from the data collected from a set of interviewees, it is advantageous to discover whether or not this emerging theory is a new or prevalent theory among the theoreticians and practitioners of the field. Theories and insights that need further exploration can emerge at any point in the research process.

Scope of Review

As one approaches this point in their research process, typically the subject of the scope of one's study arises as a key question from the students. In an earlier chapter we discussed how limiting one's context can limit the scope of the study. In this chapter, we also have seen how identifying one's domain can limit scope. Establishing expectations for the literature view is a third place in which to either narrow or expand one's scope of study. Students will often ask, "How many references do I need to include?" My answer to this question depends upon whether the student is pursuing an advanced academic degree (such as a PhD or M.S.) or an advanced practitioner's degree (such as an MBA or DMin). For the academic degree, I suggest that the student narrow the topic as much as possible, but then for the literature review, identify every primary source associated with that topic. The best way to ensure that all salient resources have been identified is to review the bibliographies of the seminal resources identified on the topic. In contrast, for a practitioner degree, I suggest that the student identify a particular domain for the study and then choose a foundational author or group of authors from which to draw. The former review can easily result in a review of over one hundred resources, while the latter may result in a review of twenty to twenty-five resources. In my mind, it is important for every religious study to have a sound theoretical and theological base, but not every study should have the same scope, based upon the goals of the degree being pursued.

Chapter 6

The Research Design

THE CONCEPT

THE OTHER DAY I spent about an hour reviewing the expectations and logistics for an upcoming consultation that I would be conducting in the near future. We covered the desired outcomes, the assumptions of the participants, the tension and resistance points, the draft of the process, and the many other key points during the consultation.

Near the end of the conversation, my contact person said, "I think that should cover it."

"I do have one more question," I added.

"What is that?"

"Where will the event be held?"

After a little laughter . . . "I guess that would be an important piece of information to have as well!"

The research design for the project is the roadmap for the project. The design should contain all of the necessary pieces of information for both the researcher and the reader so that all persons involved clearly know what is expected. The design directs the researcher toward the destination. Just as it would be disastrous to land via air travel in a particular city for a meeting but not know where the meeting is being held, so must the research design contain all of the critical morsels of information to channel the researcher toward the endpoint.

While the research design should contain all of the necessary details to guide the researcher to an anticipated conclusion, the research question also has its own comprehensive quality.

The true test of a research question is, "If I answer this question, will I have completed my project?"

If the answer to that question is yes, then the researcher has composed a solid research question. If, on the other hand, the answer is, "no," then the researcher has more work to do. The research design and research question work hand in hand and actually contain the same information. The research design simply provides more detail than the research question itself.

One way to conceive of the two pieces is to consider the research question as supplying the panoptic view of point A and point B and the research design as supplying the anticipated route. The research question should provide the beginning point, context, and desired destination. A thorough accompanying research design will then answer the salient questions for the conductors and observers of the project, namely: who, what, where, when, and how. The "why" should be fully known at this point, having arose out of the researcher's passion, interests, background, and skills, as covered in previous chapters. Other details, however, should be fully nuanced. This is the time for the researcher to engage his or her pointillistic side and embrace the minute details of the research design.

COMPONENTS OF THE RESEARCH DESIGN

Who

The research design should cover who is involved in the project. This includes not only those who will conduct the research, but also those potentially affected by the study. Be sure to clearly define your own leadership role in the project, carefully describing how you will be involved in each phase of the study. Some researchers serve as the data collector, while others do not. Some work with a steering group or leadership team. If that pathway is chosen, plainly define your role with the team. Beyond the role of the researcher, identification of the stakeholders can extend quite broadly. The following categories listed below can aid the researcher in surfacing the stakeholders of a project.

1. The primary researchers

2. The leadership team

3. People involved in collecting the data

4. People who will be asked to provide data

5. People interested in the results of the project

6. Supporters of the project and its ministries

7. People potentially affected by the outcomes of the project

8. Faculty members and administrators of the institution

9. People who serve as gatekeepers for the project or its participants

10. Family members of participants

11. Organizations to which the participants belong

12. People potentially affected by the unintended consequences of the project

The group of persons listed above that should be given the most attention when describing the stakeholders are those who will be supplying data to the researcher. This group will be covered in more detail in subsequent chapters on data collection and data presentation, but for now, know that it will be important to provide a thick description of the persons or groups serving as the pool or sample for the project. For other groups in the list above, the most important aspect of the research design is to acknowledge their connection to the project prior to collecting the data and to gain necessary permissions from relevant individuals and organizations affected by the study. Simply recognizing a certain group sends a signal to the reader that the researcher is aware that the study might affect these persons, even though the extent of the affect may not be known a priori.

What

Prior chapters detailed the conducting of the literature and theological review related to the study. There is no need for me to cover that ground again, nor you if you are drafting your prospectus as you progress through the book. Simply remind the reader of the big picture of your study, such as demonstrating how the study will advance the overall mission of your particular context. Beyond the big picture, the "what" of the research question and research design should then drill down into the words contained in your research question. Because the research question contains the beginning and endpoints of the study, it is important to define every term in the research question. Some words that may seem clear-cut to their author, may have multiple interpretations to the reader. For example, if your study involves a mentoring role, it will be imperative to carefully define what is meant by the word, "mentoring." Would "coaching," or "guiding" be a more definitive term? Do you have the right word for each of the concepts that you are portraying? Words like "impact," "leader," "change agent," "implementation," and "formation" can be especially difficult to define and should be given great care. Every word in the research question potentially takes the reader to a new signpost in the study and thus should be carefully chosen and defined.

When

The research design should also contain a detailed timeline of the study that begins with when you will form the conceptual framework and ends with the anticipated dissemination of the findings. A number of additional questions should be answered by the timeline. When will you conduct the literature review? When will you gain permission to survey the respondents? When will the data be gathered and in what stages? When will you analyze the data? When will you write the findings and present it to the official committee? When will the data be disseminated beyond the committee? Timelines can be adjusted, but it is important to set forth an initial detailed timeline in the research design. The process of specifying a timeline is important not only for chronological reasons, but this activity also serves as an excellent tool for evaluating the realistic nature as well as the scope of the study.

Where

Where the study will be conducted primarily applies to three areas, namely the context, the pool, and the sample. Context is the largest environment, followed by pool and then the sample. Context refers to the groups or persons from which you are seeking to discover something or develop a particular aspect of the organization (as referenced in chapter 3). The pool is descriptive of the specific persons or groups from which you are drawing data, either qualitatively or quantitatively. In the research design it is important not only to name the pool, but also to demarcate how you will gain access to those persons. Does the organization have an official board that can be approached to gain access to these persons? Are they a group of leaders who may require some incentive in order for them to respond in greater numbers? Carefully define the pool of persons or groups to which you will seek to gain access. Finally, the sample is the smaller group from within the pool that you plan to actually interview or survey.

To illustrate context, pool, and sample, imagine that you are conducting a study that seeks to develop the sense of community of a single congregation. Perhaps you are the pastor of that congregation. The context of your study would be the congregation that you are serving. The context should be described prior to detailing your research design and would include the salient relationships, patterns, demographics, geography, functions, and practices of your congregation. The pool of your congregation typically would be the membership of your congregation or perhaps the active participants. The sample would be those persons who would receive the actual invitation to participate in a data gathering process such as a survey or focus group or be interviewed by a member of the research team. In some cases, the sample and pool may be the same group, which would be true in this instance if you chose to survey the entire membership of the congregation.

How

Answering the "how" question of the research design largely involves diagraming and sequencing the data collection process in a manner that arrives at the desired final product of the study. In order to illustrate this part of the research design, we must dig a little deeper into some classic research designs.

DIGGING DEEPER

To dig deeper into the formulation of the research design, several classic research designs will be detailed below. Your specific study may follow one of the designs characterized below; it may involve a hybrid of two or more of the designs; or the study may dictate its own design unique to and dictated by the parameters of your study.

Experimental Designs

Experimental designs are often used in development studies. They follow a classic sequencing of:

- Assessing an aspect of the organization.
- Providing an intervention designed to improve that aspect of the organization.
- Assessing the organization again to see if the desired impact occurred.

While classic experimental designs, "attempt to isolate and control every relevant condition which determines the events investigated and then observes the effects when the conditions are manipulated,"[1] this is simply not practical in a congregational or religious institution setting. Rather, in religious studies, the emphasis is placed upon an intervention or set of interventions that are believed to be able to produce the preferred improvement. It would be unethical in most situations to deny a control group an intervention that they desired to receive in a religious setting. It is also nearly impossible to control several aspects of an organization while seeking to influence a single aspect. Rather than controlling for other factors, the researcher details their context and leaves the burden upon the reader whether or not to draw practical applications from the study.

A sample research question from a former student that is patterned after an experimental design is shown below:

"Will a new worship service paired with a comprehensive plan including the articulation of core values and discipleship pathways change the generational and ethnic make-up and result in a change of identity of First Church?"

---

1. Walliman, *Research Methods*, 11.

Typically, the desired final product of an experimental study is an activity or set of activities that have been shown to lead to or not lead to an improved condition in the organization. Seldom is a clean outcome achieved with this type of design. Rather, the final product often includes some examples of what worked and what did not work with suggestions of future experiments.

Comparative Designs

Comparative designs seek to compare how a particular decision, practice, theory, etc., is being applied or is affecting two or more differing contexts. These designs have been used to study the effect of ecclesiastical decisions on different judicatories, the responses of different communities to the same communal circumstances, and the health and vitality of congregations in differing environments. The desired final product of such a study is a set of insights into the similarities and differences of the two contexts. Such a study can also look at how similar communities respond differently to similar circumstances as in the sample research question below:

"How did the three congregational communities of A, B, and C respond differently to the severe economic conditions present in all three communities in ways that led to differing outcomes of self-image, core values, and community transformation?"

Action Research Designs

The desired end product of an action learning design is very similar to the desired end product of an experimental design, but an action learning approach may allow the researcher to attend to different issues than the experimental design. Action research designs follow a specific pattern of:[2]

- Identification of a problem or set of issues
- Reflection on the problem
- Action and experimentation
- Identifying learnings
- Repeating the cycle

Action learning cycles have been applied to a variety of contexts and are particularly well suited for addressing adaptive issues as opposed to technical issues. Technical issues are those for which there is a known solution, while adaptive issues are those that require some work in order to surface a better understanding of the problem itself according to Ronald Heifitz, who popularized the notion of adaptive change.[3] Researchers who enjoy tackling these types of problems may consider an action learning approach to design their respective study. A sample research question from a former student of mine that employed action learnings cycles is shown below:

"Can overseas mission act as a changing agent in building and shaping future leaders for First Church?"

Descriptive Designs

Descriptive designs are often used to study new phenomena. These designs have been used in recent years to study megachurches, storefront congregations, and first and second generation immigrant congregations. The goal of a descriptive study is to describe

2. Marquardt, *Action Learning*, 28.

3. Heifitz et al., *Practice of Adaptive Leadership*, 19.

the organization from an insider's perspective. These studies typically draw heavily from qualitative as opposed to quantitative methods, and might involve several concurrent methods of study. For example, the researcher may be conducting observations, focus groups, interviews, and brief surveys concurrently in order to gather the desired data. The final product of this type of study is a narrative that serves as a window into the emerging type of organization studied. A good resource that contains summaries of several of these types of studies is Stephen Warner's book entitled *Gatherings in Diaspora.*

Naturalistic Inquiry

Closely related to the descriptive design is a design coined by Egon Guba entitled, "Naturalistic Inquiry."[4] While the desired end product is the same as a descriptive study, naturalistic inquiry espouses the conducting of several iterations of data collecting in order get closer to the real picture of what is happening as the study progresses. Each iteration of data collection is informed by previous iterations as the researcher draws closer and closer to the desired product. The accuracy of the product is determined by the key stakeholders whose input is sought after each round of data collecting. A sample research question utilizing naturalistic inquiry from a former student of mine is shown below:

"How is worship redefined within a missional praxis for a 17–27 year old community?"

Historical Research Designs

Historical research designs delve into previous events, and are aimed at uncovering processes, decisions, and actions rather than a narrative of the newly emerging sociology. Historical designs

4. Guba, *Naturalistic Inquiry.*

simply ask, "What really happened?" Historical designs have been applied to studies of war and the responses to it, racial/ethnic issues, and dramatic events such as the Holocaust. A colleague of mine is currently working on a historical project seeking to uncover a denominational response to the Holocaust. A sample research question utilizing a historical design from a former student of mine is shown below:

"What is the role of fictive kinship on faith
and longevity among dually diagnosed men
institutionalized for more than twenty years at
x institution?"

Grounded Theory Designs

The desired end product of a grounded theory approach is to surface a working theory for the topic of the study. "Grounded theory takes the approach of collecting data in order to evolve a theory rather than test or refine an existing one."[5] The design may employ several sequential steps or employ simultaneous mixed methods, but the goal is to develop insight into the topic that may serve others in similar contexts. A sample research question from a former student of mine is shown below:

"How can the Pan-Methodist congregations in
the common neighborhood of an urban setting, utilize their common Wesleyan heritage
for the purpose of establishing a mission that
impacts the community?"

Although some of the classic research designs seem very taut, it is important to keep in mind that the research design is merely

5. Walliman, *Research Methods*, 102.

an anticipated route. Anticipated routes often contain detours, ditches, and delays and some routes are more prone to cunctations than others.

CLASSROOM EXERCISES

One of the most appreciated aspects of my research design classes is giving every student the opportunity to receive feedback on their proposed research question. I typically provide students with this opportunity midway through the course and once again at the end of the course. Over the years, students have been amazed at the helpful feedback that they receive during this process. Some drafts of research questions change very little, and others hardly resemble the original draft once the feedback is received, but all appreciate the opportunity.

Depending upon class size, I allot a total of twenty to thirty minutes per person for the students to post their research question on a wipe-off or smart board, allowing sixty seconds of verbal commentary from the presenter, and then invite their peers to offer feedback. I enter into the discussion as well, but most students are energized to both receive and provide feedback for one another as they seek to apply their newly developed research tools.

If this practice is offered, it is important not to allow too much verbal commentary from the presenter. Remember the key concept in this chapter is that the test of a quality research question is that it can stand on its own. If one can envision being through with the research process once the research question has been answered, then it is a quality research question. Thus, if too much commentary is required by the student to explain the verbiage of the research question, then the research question needs more precision. In my experience, it is better when the need for more precise language surfaces from others in the classroom discussion rather than having the student realize that he or she has consumed too much time in the preliminary explanation.

ADDITIONAL THOUGHTS AND RESOURCES

In this chapter, we learned that the desired product of the research study prescribes the type of research design appropriate to one's study. If the researcher is seeking to surface a process from which others can learn, then an experimental or action learning approach is probably best suited for this researcher. If the researcher desires a product that focuses on the influence of context in research, then a comparative design might be most appropriate. Historical and descriptive designs are most appropriate for researches who want to add to the knowledge base for future practitioners. Naturalistic inquiry is appropriate for researchers drawn toward the phenomenon of early adopters. Finally, researchers who love studying theory and who desire to create their own theory might consider the grounded theory approach.

It is important to continue to keep in mind that every decision made will affect subsequent parts of the funneling process, lest the funnel turn into a farrago. Each part of the funnel powers the next as the key question in each part of the funnel continues to drive the ensuing process. By now, readers who have worked through the funnel approach to designing religious research should recognize that they are more than halfway through the process. We turn our attention next toward an often misunderstood and extremely important part of the research process: the selection of quantitative and qualitative methods for pursuing one's research question. This is where the fears of many persons new to the research process have accumulated over the years. Timorous readers will be delighted to discover that the methodology of the study is driven by the culmination of all previous decisions, making the selection of methods quite simple and straightforward.

Chapter 7

Quantitative and Qualitative

THE CONCEPT

Whenever an anonymous pollster phones our house the typical response from the person answering the phone is to hand the phone to me with the words, "It's for you." While I realize that there is a high probability that the person about to ask me questions on the other end of the phone did not actually write the questions being asked, this never deters me from engaging the questioner in a dialogue about the questions themselves. I am happy to respond to questions that are clear and concise and do not lead the respondent toward a particular response. The problem is that I do not get many of those. Sometimes the first couple of questions are straightforward and I reply with a direct response. But by the third or fourth question, I am usually responding with, "Have you ever considered that beginning a question with, 'Don't you think that . . .' might tend to influence the person to respond in a certain way?" I am also not immune to asking the pollster questions of my own such as, "Do you know what percentage of persons are prone to give socially desired responses?" At some point, I usually point out that not only will their data be skewed by the questions being asked, but that the very format being used to ask these types of questions is probably not the best choice of formats. I seldom get the opportunity to hang up on a pollster. Most of them initiate that courtesy themselves.

BREADTH OR DEPTH

When choosing tools to gather data, it is very important to choose the correct tool for the task. Just as it is not appropriate to use a meat thermometer to open a can of vegetables, it is essential to select the correct tool in order to gather the desired type of data. Choosing the correct tool involves first answering a basic question, "Are you more interested in expanding the breadth of what you currently know, or are you more interested in knowing which parts of what you already know are substantive?" Prior to gathering data, researchers must answer the "breadth or depth question" for themselves.

Qualitative research adds to the breadth of your knowledge base while quantitative research deepens the substance of what you already know.

Qualitative research adds breadth to your content. Quantitative research adds depth to your substance.

Imagine that you are studying persons' motivations for getting more involved in a congregation. If you have a few hunches, but greatly desire to expand your hunches, then a qualitative approach to broaden your content would seem most appropriate. But, on the other hand, if you feel like you already have a handle on all of the reasons why people get more involved in a congregation, but would love to know which of those reasons are the most substantive, a quantitative approach would serve you better. Persons, including telephone pollsters, opinion gatherers, MBA students, Doctor of Ministry students, and PhD students frequently use the wrong tool for the type of data desired. Unfortunately, much of this contraband data even finds its way into the newspapers that we peruse and the articles that we share at the water cooler. Studies are regularly published that show signs of incorporating the wrong technique for the desired outcome. Let's explore further the notion of employing the right research tool in the correct setting.

CHOOSING THE BEST APPROACH

One of the most prevalent mistakes is to use a quantitative method when seeking to gather more content. With the explosion of the internet, persons have fallen in love with surveys. Surveys are a quantitative tool, engineered to provide substance to what the researcher already knows. Surveys work very well when persons can check boxes, rank choices, and click on ratings. Surveys do not work very well when they ask open-ended questions. Respondents regularly skip those questions and when they do answer them, they often delve into an area unintended by the researcher. If these types of open-ended questions were being asked in a face to face interview, the researcher could clarify the question being asked, add follow-up questions, and ensure that the query gathers the type of data desired.

People gathering data also employ qualitative techniques in quantitative situations. Sometimes interviewers will ask a person a series of closed-ended questions, the kind of questions that someone could respond to by checking boxes. Naturally the interviewee becomes frustrated and often begins responding with material that is completely off topic, frustrating the interviewer as well. Checking boxes is for surveys not interviews.

Have you ever heard that someone is a mile wide and an inch deep? The same applies to qualitative studies that have plenty of content, but have no idea what is most substantive. Sometimes researchers go back and try to count how many times a particular word was used in an effort to quantify their qualitative data. Suppose that the question dealt with stress. If a person uses the word "anxiety" instead of stress, should that count? What about the word "overwhelmed"? If you plan to count your responses, use a quantitative method. If you plan to create themes out of what you gather, use a qualitative method. Quantitative research seeks to test out something about a group, such as that the group has a certain level of stress or that group A and group B are substantively different in their stress levels. Qualitative research seeks to uncover information that you do not already have.

DIGGING IN

Selecting the best approach for your study involves deciding upon a qualitative or quantitative approach. A qualitative approach adds breadth to your base. A quantitative approach adds depth to your substance. Within each of these two approaches, lie many options for collecting data. These options are called procedures. Below we will briefly explore two of the more common procedures for each of these approaches. In the next chapter, we will delve further into the realm of procedures by discussing best practices associated with research procedures.

QUALITATIVE PROCEDURES

Interviews

Conducting interviews is the most widespread qualitative procedure. Interviews are best done face to face, but can also be conducted over the phone. Because interviews fall into the qualitative side of research methodology, they should be conducted when the researcher desires to add to their knowledge base. Interviews provide an excellent opportunity to sit across from someone and pose open-ended questions and record the content while also noting the accompanying emotions, facial expressions, and energy.

While all interviews pose open-ended questions, there are different levels of open-endedness. Unstructured interviews pose the greatest opportunity for the researcher to delve into brand new territory with the respondent, but may also take the interviewer beyond the scope of the intended project. Structured interviews typically contain twelve to fifteen questions for a sixty to ninety minute interview. They still allow the respondent to name new content, but within a specified set of parameters set by the scope of the project. Semi-structured interviews are, as expected, somewhere in the middle. I sometimes suggest that these interviews allow for the chasing of a particular rabbit, but do not allow rabbits to become the main topic.

Focus Groups

Focus groups allow the respondent to interview a group of persons as if the group were a single person. Every existing group has its own personality, culture, and set of behavioral characteristics. While newly formed groups (which comprise the best focus groups) have no history, they can typically open a window to the characteristics of the group that they are intended to represent. Conducting focus groups is a way of gathering the group perspective on a particular issue among a particular group of people. In order to surface the group perspective, ask, "Does this emerging view reflect the perspectives of everyone here?" The best focus groups contain two elements, namely anonymity and commonality. In other words, the best focus groups are formed by persons that do not know one another, but all have a certain characteristic in common. The desired commonalities may be demographic such as age, ethnicity, tenure, etc., or they may be ideological or theological. The desired commonality should arise out of the researcher's determination of what type of group can supply the richest new content.

QUANTITATIVE PROCEDURES

Surveys

Because a survey is a quantitative procedure, it is should be used to enhance the substance of what one already knows rather than seeking to expand the knowledge base. Surveys and instruments are the two most common forms of quantitative tools. The difference between a survey and an instrument is that surveys cover multiple topics, while instruments cover a single topic. The best surveys use a variety of questioning techniques, such as asking the respondent to rate a series of items on a number scale, put a collection of items in rank order, or check categorical boxes. Each of these different types of survey questions deals with a different type of quantitative data which will be explained below.

Nominal or categorical data refers to choices that have no hierarchy or order to them, such as a person's gender or ethnicity or whether a person grew up spending more time indoors or outdoors or grew up eating more pie or cake. Ordinal data refers to an ordered set of choices such as asking a person to rank the set of movies nominated for best picture from most enjoyed to least enjoyed. If, however, the person was asked to rate each picture on a certain quality, such as how humorous or realistic the movie was on a scale of 1–10, this would serve as an example of interval data. These three types of data can be illustrated by imagining persons competing in a race. The demographics of the competing persons would be considered nominal data. Whether the person finishes first, second, third, etc. would be considered ordinal data, and the time that each person took to complete the race would be considered interval data.

The majority of questions contained in a survey should ask the respondents to rate, rank, or check boxes. It is acceptable to add a few open-ended questions to a survey as well, but this is done more to satisfy the need of certain respondents to give more detailed information about a particular item. Remember that if the main desire of the researcher is to add content, then a qualitative rather than quantitative approach should be utilized.

Instruments

An instrument covers a single topic rather than multiple topics. It may seem odd to design an entire document to cover a single topic, but when that topic is so salient to one's study, then this is clearly the preferred procedure. Instruments are often employed by a researcher who is trying to show that either a particular aspect of an organization has improved, or that a group of persons have changed their attitudes or grown in a particular way. For instance, if someone is seeking to enhance the sense of community within a congregation, and the researcher desires to collect data about sense of community, then an instrument would seem to be the best procedure to apply in this situation.

The most common forms of instruments actually contain a series of statements rather than questions. The respondent is usually asked to express his or her level of agreement with each statement. It is not uncommon for an instrument to contain a few dimensions within an overall construct or aspect. For instance, an instrument seeking to measure a sense of community, might actually measure the willingness of the persons in that organization to support one another, their desire to spend time with one another, and the respect that they hold for one another. In this instance, the instrument would produce a separate measure for support, time, and respect. Each of these dimensions would then be aggregated to produce a "score" or measure for sense of community.

The advantage of using a data gathering procedure that focuses on a single topic is that the validity and reliability of that instrument can be increased by doing so. Validity refers to the ability of an instrument to measure what it is intended to measure. Reliability refers to the ability of the instrument to measure that intent consistently over time. A validated instrument would demonstrate the relationship between each of the dimensions of the survey and the aggregate or whole being measured. The validity might also show the relationship of the particular variable to other variables, such as the relationship between this measure of sense of community and other measures of community or belonging. Reliability data from the instrument would establish a consistent pattern of responses over time.

Different Approaches Imply Different Methods

Because quantitative and qualitative approaches have very different aims, they also call for very different methods. For instance, qualitative approaches need open-ended questions to surface new content. Because substance rather than content is the aim of quantitative approaches, closed-ended questions are employed. The data gathered from open-ended questions leads to the formation of patterns and themes, while the data gathered from closed-ended questions leads to the development of tables and charts. Tables allow for the easy comparison of groups such as pretest and posttest groups, demographic groups, and sampling and population groups. The data from qualitative and quantitative seeds looks different. Qualitative research gathers words, images, and concepts. Quantitative research gathers numbers. Numbers are then analyzed statistically, while words, images, and concepts are analyzed organically producing lists, cycles, diagrams, quadrants, etc.

Perhaps the most difficult concept to grasp regarding differences between qualitative and quantitative research is that the persons who supply quantitative data are chosen randomly or systematically, while the persons who respond in qualitative approaches are chosen purposely. This is because the burden of rigor differs for the two approaches. In quantitative approaches, the researcher must show that the substance obtained was not influenced by the particular persons who supplied the data, while just the opposite is true of qualitative research. Since qualitative research is aimed at expanding content, subjects perceived as being able to supply the richest, and broadest new content are chosen purposely.

CLASSROOM EXERCISE

Something that has not yet been named in this chapter, but that is pivotal at this point in the process is that the researcher must continue to follow the funnel approach at this stage of the research process. The research design that has been constructed thus far in the process must drive the methodology selection. Perhaps the best way to illustrate this imperative is with a classroom exercise.

Qualitative research and quantitative research have very different aims. Qualitative research seeks to broaden one's research base, while quantitative research seeks to deepen one's substantive claims. Whether or not the researcher is targeting breadth or depth should have already been determined at this point by the words of the research question. Certain words of a research question point toward breadth, while other words point toward depth. The ultimate determination of whether breadth or depth is needed is contained in the full-scale research design, but well-crafted research questions give strong clues to whether the research question is calling for a qualitative or quantitative approach to the study.

As a classroom exercise, or as a reading exercise, read through the list of verbs below. Beside each verb, write in the blank (or shout out in the classroom setting) whether you think that the particular verb is pointing toward a qualitative or quantitative approach. Suggested answers are shown at the end of this chapter.

1. Explore _____

2. Compare _____

3. Contrast _____

4. Surface _____

5. Create _____

6. Determine _____

7. Enhance _____

8. Motivate _____

9. Clarify _____

10. Build _____

11. Study _____

12. Transform _____

13. Influence _____

14. Test _____

15. Review _____

Words like explore, enhance, and influence point us toward a more qualitative approach because they signal the expansion of content. Whereas words like compare, clarify, and test signal the desire for depth of substance rather than breadth of content because they look for relationships among what already exists. This exercise can serve as an effective illustration of how the research design and research question can spell out the next step in the research funneling process. The verbs of the research question should drive the employed methodology.

While teaching this research design class to a group of Doctor of Ministry students in the country of Myanmar, I had realized that the dynamics of this hierarchical and communal culture did not allow for individual responses in front of others, so I created two different ways for students to respond that produced two different sounds, both of which could be produced anonymously under the desk. One sound was made by thumping on the bottom side of the wooden desk, while the other sound was made by clanging two pieces of metal utensils together. Depending upon the culture of your group, it may be necessary to create an alternative to shouting out due to the fact that some cultures do not lend themselves to that form of behavior.

ADDITIONAL THOUGHTS AND RESOURCES

Suggested Resources

Many studies call for a mixture of quantitative and qualitative approaches rather than a single approach. These types of studies are called "mixed methods" approaches and are quite common in research.[1] For instance, a researcher may gather data qualitatively through exploratory interviews in order to design a better quantitative survey. A researcher may use parallel approaches of

1. Creswell and Clark, *Designing and Conducting*, 62–75.

qualitative and quantitative in order to determine if the two approaches lead to similar outcomes. This process is known as triangulation. The researcher may gather data quantitatively and then employ the use of an explanatory focus group to help verify or elicit stronger meaning of the gathered data. Finally, one approach may simply be embedded within the other approach.

In the notes above, only two procedures were discussed for each of the two approaches to qualitative and quantitative research. While the elucidated procedures are two of the more common research procedures for each approach, there are many more procedures that can be utilized to address the need for gathering data. Most of the popularized procedures address the need for increasing one's knowledge base rather than adding depth of substance and thus, fall into the qualitative methods pool. Additional quantitative procedures exist, but may require the assistance of a person trained to administer and interpret that particular tool. Most of the newer qualitative approaches suggest a very specific set of processes to follow in order to obtain the desired outcome as illustrated below.

Appreciate inquiry has been used by numerous organizations to surface data about what an organization looks like when it is at its best throughout history. An "Appreciative-Inquiry Summit" is an excellent process for gathering a large amount of data in a short period of time. The summit is based upon the principles of (1) discovery, (2) dream, (3) design, and (4) destiny.[2] A similar approach is called, "Asset Mapping." This approach seeks to surface new content of the organization in the area of assets. Asset mapping surfaces the physical, individual, associational, institutional, and economic assets of an organization and then connects the dots among them.[3] The downside of these two approaches is that they tend to ignore the current challenges of the organization. Ronald Heifitz has developed a process specifically designed to surface the deepest challenge within an organization which he terms "adaptive challenges." Heifitz claims, "The single most important skill

2. Ludema, *Appreciative Inquiry Summit*, 10.
3. Snow, *Power of Asset Mapping*, 4–5.

and most undervalued capacity for exercising adaptive leadership is diagnosis."[4] His latest book on adaptive change contains several qualitative diagnostic procedures that show the reader how to gather data related to the deepest challenges of the organization.[5] A tool for gathering data through conversation is known as "Conversation Café," or "World Café." The key to these conversations are the questions posed for discussion. Brown and Isaacs do an excellent job of showing how to generate questions that:[6]

1. Are simple and clear

2. Are thought provoking

3. Generate energy

4. Focus inquiry

5. Surface assumptions

6. Open new possibilities

There are also many qualitative resources designed to create a vision or strategic plan for the organization such as "Future Search"[7] and "Whole-Scale Change."[8] There are a multitude of resources for data collecting. Potential resources should be thoroughly vetted prior to using them in a study. The key is to select the most appropriate resource for the desired task. In the beginning stages of the funnel, the emphasis was upon creativity and exploration. Now that the research design has been drafted, the emphasis is much more upon aligning the chosen procedures with the chosen design. As Creswell and Clark point out, "the methods should match the problem."[9]

---

4. Heifitz et al., *Practice of Adaptive Leadership*, 7.

5. Ibid.

6. Brown and Isaacs, *World Café*, 165.

7. Weisbord and Janoff, *Future Search*.

8. Dannemiller, *Whole-Scale Change*.

9. Creswell and Clark, *Designing and Conducting*, 32.

Suggested Answers to the Classroom Exercise

1. Explore (ql): Explorations signal new territories.

2. Compare (qt): Comparisons are usually made of existing data and known circumstances.

3. Contrast (qt): You cannot contrast what you do not already know.

4. Surface (ql): Surfacing makes previously hidden information visible.

5. Create (ql): Creating adds new ideas or identifies new perspectives.

6. Determine (qt): Determinations are made to see if something is present or not present, but the investigated item is usually known.

7. Enhance (ql): Enhancements lead to an expansion of knowledge.

8. Motivate (ql): Most research questions that include words like motivate and recruit are looking for new ways to accomplish those tasks.

9. Clarify (qt): You cannot clarify a blank page.

10. Build (qt): Most builders already have their building materials.

11. Study (ql): We study to learn new content.

12. Transform (ql): Transforming usually involves more than rearranging deck chairs.

13. Influence (ql): Discovering this almost always covers news ground.

14. Test (qt): A grade is based upon what one already knows.

15. Review (qt): We can only review what is already there.

Chapter 8

Data Collecting Procedures

THE CONCEPT

Research is Like Planting Seeds

IN THE LAST CHAPTER we discussed the difference between qualitative and quantitative research. Remember that qualitative research is intended to add breadth to your base of knowledge, while quantitative research is intended to add substance to the knowledge that you already have. In many ways, gathering data is like planting seeds. Each question that is posed is like planting a seed in the soil with the intent of seeing how it responds to its new environment. It is important to know the environment and to know what is being planted. Obviously, it is impossible to grow corn from carrot seeds or beets from avocado seeds. It is important to plant the type of seed which will result in a harvesting of the kind of information desired from the respondent. While there are a multitude of different questions that can be asked of respondents, quantitative and qualitative questions each have their own distinct type of seeds.

Quantitative research is like planting grains of corn. There is very little doubt where the stalks will grow. Rather, the mystery lies in how many grains will grow on each ear and which stalks will be the most substantive. Qualitative research on the other hand, is like planting pumpkins. Once pumpkin seeds are planted, they may show up anywhere in the garden, occupying new spaces and new places. The vines may even extend beyond the boundaries of what is intended to be the garden! The fruit also comes in varying sizes, unlike corn. The mystery of a pumpkin seed is horizontal while the mystery of a corn seed is vertical.

Whether one is planting pumpkins or corn, careful preparation of the soil will benefit the harvesting. In the case of qualitative procedures, a proper harvest implies a rich yield of new content. In the case of quantitative procedures, a proper harvest implies accuracy of outcomes.

Asking questions of potential respondents is like planting seeds; the soil must be prepared, proper planting must take place, and care must be given to the harvest.

DIGGING IN

Preparing the Soil

The main concern in preparing the soil is to reduce potential response bias. Bias occurs when there is a distortion in the results of the data, not accounted for by the researcher. In preparing the soil for planting seeds or asking questions of potential respondents, the researcher must take several aspects into consideration, including the group being sampled, the consent of the respondents, and the validity of the questions themselves.

Population and Sample Size for Quantitative Procedures

Population and sample are two key terms that should be understood by every researcher. They apply in different ways to qualitative and quantitative research. First, let's cover quantitative research. The context of a study is the group of people or set of organizations that you identify with as a researcher. The population is the set of all possible respondents in that group. The sample is the list of people who actually receive the questions. To carry the agriculture analogy one step further, the population of a research study would be one's entire farm and the sample would be the section of acreage designated for planting a specific crop. Different crops may be planted on the same farm.

The population is always larger or equal to the size of the sample. In some cases, the sample and population may be the same group. For instance, it is not unusual in religious studies for a single congregation to serve as the population for the study and then for the survey to include every member or every active participant of that congregation. This can be achieved by mailing the survey or distributing the online link to every member or participant of the congregation. The problem with this method is that it is not face to face and will yield a much lower response rate than distributing the survey face to face. If you are doing research on a single congregation and plan to distribute a survey to them, I strongly recommend that you distribute the survey following worship. This is a widely acceptable practice in congregational research. Technically, the set of respondents will be a sampling of worshipers who happen to attend on a given Sunday, but this method is preferable to mailing out a survey and receiving fewer total responses. Remember that the key consideration in research sampling is bias. The likelihood of the group of people who choose not to respond to a mailed survey being different from the group of people who choose to respond to the survey is probably higher than the group of people who happen to show up on any given Sunday being radically different from a group people who happen to attend on any other given Sunday. When distributing a face to face survey, try to reduce the survey to set of questions that can be completed in ten to fifteen minutes. If you have multiple worship services, distribute the same survey following every worship service that occurs on the same weekend. Advertise the survey in advance and distribute and collect it on the advertised date.

Sometimes, it is simply not practical to distribute a survey face to face due to the size of the congregation or the desire to collect a large data set from several organizations. In this case, it is best to draw either a systematic or a random sample from the population. Begin by putting the population of potential respondents in some type of order, typically arranging them alphabetically. A systematic sample draws every tenth, twentieth, thirtieth, etc. name from a larger population. A systematic sample is not random, because every person in the population does not have an equal opportunity of being selected for the sample, for instance multiple family members with the same last name would never be chosen. A random (and less biased) sample could be drawn from a large population by assigning every person a number and then drawing random numbers from a computer generated app or program and adding each person whose number is drawn to the sample until the desired number of respondents is reached.

Regardless of whether one surveys the entire population or draws from a smaller sample within the larger population, not every person will respond to the survey. Distributing the survey or instrument face to face to the entire population, such as following a worship service, carries the greatest chance of achieving a high response rate. Beyond that format, response rates are greatly overrated. Acceptable response rates have been quoted from 10 to 90 percent.[1] Further research suggests that seeking to increase response rate does little to reduce response bias.[2] This makes intuitive sense. Suppose that you achieve a laudatory response rate of 80 percent, but later discover that the group of 20 percent who did not response differ greatly from the group of 80 percent who did respond. Is the 20 percent enough skew the data? Absolutely. Rather than tripling one's effort to achieve a high response rate, put that time into designing questions that are clear and sound and fit the environment of your population to reduce potential response bias. In short, continue to prepare the soil.

1. Radwin, "High Response Rates," 8.
2. Ibid., 9.

In addition to response rate, the researcher must consider the size of the sample if the sample needs to be smaller than the entire population. Once again, there are many misconceptions about sample size. Please know, that there is no such thing as a standard portion of a population in which to sample. "A sample of 150 people will describe a population of 15,000 or 15 million with virtually the same degree of accuracy, assuming all other aspects of the sampling design and sampling procedures were the same."[3] As it turns out, 150 is a good number for a sample size. A number less than that will increase one's sampling error (a concept that we will discuss in the next chapter), and a number larger than that is simply not needed in most cases. "Precision increases rather steadily up to sample sizes of 150 . . . After that point, there is a much more modest gain to increasing sample size."[4] The ultimate determination of sample size should be made on the basis of having enough people in each of the subgroups that the researcher desires to compare. Most religious studies compare minimal subgroupings, in which case the 150 number will more than suffice. In conclusion, unless your population is considerably larger than 150, try to survey your respondents in a face to face setting or survey the entire population electronically if face to face is not an option.

Population and Sample Size for Qualitative Procedures

Selecting a group of people for a focus group or selecting a group of people to interview is much less complicated than choosing a sample for a survey or instrument. That is because qualitative and quantitative methods have very different aims. Remember that the intent of quantitative research is to reveal what pieces of what you already know are most substantive. Thus, in quantitative research, the emphasis is placed upon selecting a group of people who are unbiased. Qualitative research has an opposite aim, namely to expand one's content, to expand what is already known. Thus, in

3. Fowler, *Survey Research Methods*, 41.

4. Ibid., 43.

qualitative research the emphasis is placed upon choosing persons who can provide the richest content. In quantitative research, the selection of respondents is random, but in qualitative research the selection of respondents is purposeful.

When selecting a group of persons to respond to open-ended qualitative questions, choose persons that you believe can give you the fullest answers to those questions. If you are interviewing persons about their sense of community, consider where differences of opinion may surface and select five to seven persons in each category to interview. For instance, it seems reasonable to think that the tenure of a congregant may influence their view of community, so interview several persons who have been participants for less than five years, several persons who have been participants for five to ten years, and several persons who have been participants for more than ten years. If you think gender may be a factor, interview several men and a several women. Consider where you expect differences to occur either from your own hunches, or better yet from your literature review, and purposely select interviewees who can supply you with helpful brand new content.

In terms of how many persons in a particular grouping to interview, six persons is a reasonable number. "When the purpose of the interviews is to create a rich description of experiences rather than to explain them, a small sample size of about six persons is appropriate."[5] In terms of the total number of interviews to conduct for a specific study, fifteen to eighteen interviews is a good number, taking into consideration that, "Saturation or redundancy of information can typically occur after 15–18 interviews."[6]

Using a Tool from an Existing Study

Bias can creep in from the selection of potential respondents, and it can also be introduced by the questions themselves. One of the best ways to avoid response bias is to use a set of questions that has

---

5. Moustakas, *Phenomenological Research*, 14.

6. Swinton and Mowat, *Practical Theology*, 145.

been tried and tested in prior studies. Believe it or not, researchers love to share questions with one another. I belong to more than one research group who regularly asks of one another, "Does any have a proven way of asking . . .?" No matter how much effort one puts into designing questions that lack bias, the best way to remove bias is to distribute the questions to a pilot group and then run a set of statistical procedures on the data to remove unwanted questions. For instance, questions that produce no range of response, such as every person responding to the question with a 1 or a 5 will yield no variability among the respondents and should be thrown out. Questions that are designed to work with a grouping of similar questions, but for some reason, work against them should be thrown out. Several other procedures will be run, but the point is that proven studies have already done this work for you.

It is entirely appropriate in research, once permission is obtained, to borrow a survey or instrument from another researcher. If you are conducting a study on building a sense of community in a congregation, as was discussed in an earlier chapter, go to the library or look online to discover a study that has already developed an instrument to measure sense of community. Or pull specific questions from a variety of published surveys that match the content that you wish to receive.

Seven Characteristics of Clear Questions

Sometimes, it is not possible to use an existing set of questions. In these cases, the researcher will need to develop their own set of questions. Below I have included seven characteristics of clearly written questions that I have refined over the years. I use these characteristics in a classroom exercise that I will discuss later in the chapter, but for now I simply present them for review. Most questions that lead to response bias break one of the rules in the seven characteristics below.

1. The question focuses upon one clear thought.
2. The wording is clear and void of ambiguous vocabulary.

3. The thought is concise with no extra words in the sentence.

4. The language avoids emotionally charged words.

5. Leading questions are avoided.

6. It draws upon language familiar to the context.

7. It avoids jargon and abbreviations.

Designing an instrument

Applying the characteristics named above to the design of an instrument or set of interview questions will go a long way to avoiding response bias. If the researcher is putting together an instrument, it may be helpful to keep in mind a few additional best practices. Remember that an instrument differs from a survey in that it measures one concept as opposed to a variety of opinions about a variety of subjects.

Most persons have experienced taking an instrument designed to measure attitudes or beliefs about a particular topic. These instruments typically use a rating scale, sometimes referred to as a "Likert scale" after its inventor Rensis Likert, that asks people to rate a series of items along a continuum from "strongly agree" to "strongly disagree." When designing this type of instrument, the "questions" are actually "statements" in which the reader is asked to rate their level of agreement or disagreement with each item. The more effective instruments typically cover a wide range of opinions associated with the topic, contain a mixture of positively and negatively worded items, and produce an overall score or measure for the topic. It can take several months and several pilot groups to validate an instrument. Some of these methods of validation are covered in the next section.

Validity

An item of research is valid if it is true to its perceived or claimed nature. Thus, a set of research questions is valid if the set measures what it purports to measure. There are many ways to demonstrate validity and it can be helpful to look for these demonstrations when reviewing potential instruments for selection. Some of these forms are highlighted below.

1. Construct validity—If an instrument measures a unique ideological or theological construct it is said to contain construct validity.

2. Concurrent validity—Concurrent validity is present if an instrument has been shown to measure outcomes that are similar to other known instruments that claim to have a similar construct or if it varies inversely with opposite constructs.

3. Content validity—If an instrument contains all of the content of a known construct, then it has content or sampling validity.

4. Predictive validity—If an instrument can predict outcomes, then it is said to have predictive or criterion-related validity.

These forms of validity typically require statistical analysis to demonstrate their containment and should be sought after when choosing an instrument.

While these forms of validity are probably beyond the scope of possibility if the reader is choosing to design one's own instrument for the study, there are a few forms of validity that can be demonstrated without the use of statistics. One of these is called face validity and derives its name from its ability to appear to measure what it claims to measure on the surface. This can be demonstrated by asking a group of stakeholders or experts to review the instrument and suggest what they think it measures. If the assessment matches the intent of the instrument, the researcher has demonstrated face validity for the instrument. The researcher can also carefully portray how the instrument was designed both technically and theoretically using the best practices contained in this chapter to obtain a level of internal validity without applying statistical procedures.

Observations

Observations are a special kind of qualitative data and may involve a building, office space, worship center, or a meeting, conference, or event. The purpose of the observation should be well-defined in the research design and could be a primary focus of the study such as observing a council meeting or worship service or the purpose could be supplemental such as providing background data for the organizational context. It is easier to observe unfamiliar settings rather than familiar settings, due to the number of defaults that swim through our eyes and minds as we seek to observe something familiar to us. If the observation of a familiar location is critical to your study, I suggest practicing on several unfamiliar places in order to hone your skills of observation. The biggest mistake that students of mine make in observations is to reference material that is only discernable apart from the observation they are making. Field notes, the recording of raw observation data, should only contain what can be seen or experienced on the day of the observation. Later, when making journal entries, it is entirely appropriate to add to one's field notes with information previously known. Ammerman, Carroll, Jackson, and McKinney's book, *Studying Congregations*, contains a very helpful checklist to review when preparing for observations, the headers of which are itemized below:[7]

1. Demographics
2. Physical Setting
3. The Event
4. Interactional Patterns
5. Verbal and Written Content
6. Meaning

---

7. Ammerman et al, *Studying Congregations*, 200–201.

Consent

When gathering data from persons, it is essential to gain consent from the person prior to asking questions of the person. For qualitative research, it is best to create a consent form, asking each potential respondent to sign the form indicating consent prior to gathering information. For quantitative studies, it is best to gain consent from the official board of the institution or congregation being surveyed. It is important to gain consent from each organization involved in the study. When making observations, consent also allows the data gatherer to take notes in a less obtrusive manner during worship services and other public events.

Consent forms should contain a brief description of the project, a list of potential benefits and limitations, a statement of confidentiality, and an account of how the data will be used. If minors are participating in the study, consent from a parent or guardian should also be obtained. Finally, if the plan is to record the session, which is strongly encouraged for qualitative research, consent to record the session should also be acquired.

Planting the Questions in the Soil

Whenever collecting data, the researcher should always spend a few moments at the beginning of the process preparing the soil. Before the respondent enters into the environment, record the name of the person, the place and time of the data collection, and any other pertinent data on the note-taking sheet. After the respondent arrives, highlight the purpose of the study and place the data gatherer in the context of the study. If you, as the researcher, are gathering the data yourself, it will also be helpful to distinguish your data gathering role from any other role that you may hold in the organization. Serving in the role of a pastor can certainly add response bias to the study when interviewing potential respondents. In some cases, depending upon the sensitivity of the topic, a pastor may want to seek the assistance of a third party to help gather data in order to avoid this bias. One former student of mine confessed that it was very difficult to gather data related to the quality of marriages in her congregation because she served as a staff member of that congregation, even though the focus of the questions was upon activities that strengthen a marriage.

Coding Qualitative Data

Because you are collecting words rather than numbers, coding is a much more difficult process with qualitative data. "Words are fatter than numbers, and usually have multiple meanings. This makes them harder to move around and work with."[8] Do not write down every word said by the respondent, but capture key words and phrases. Jot down brief summaries of what you hear. Write down something verbatim and then pause to gain a sense of the emotions of the respondent. Add impressions of emotions, facial expressions, high and low level energy moments, etc. in the margin of the paper as you write the content in the center of the paper. Later, within a few days of the data collection event, Miles and Huberman suggest adding reflective remarks to your raw notes that may include:

1. What the relationship with the respondents was like.

2. Second thoughts on the meaning of what a respondent was saying.

3. Doubts about the quality of data being recorded.

4. A new hypothesis explaining what was happening.

5. A mental note to pursue an issue further in the next contact.

6. Cross-allusions to something in another part of the data.[9]

8. Miles and Huberman, *Qualitative Data Analysis*, 54.
9. Ibid., 64–65.

CLASSROOM EXERCISE

One of the more productive exercises that I conduct in my research design classes is to create and analyze a research instrument in class. Begin by asking the class to name a potential topic for the questionnaire. The topic should be current, have a wide range of opinions associated with it, and be of general interest to the class. Previous topics have included megachurches, the authority of Scripture, ordination, clergy appointments, women in ministry, and the legalization of drugs.

Distribute a set of five 3 x 5 cards to each student and ask each student to write five questions for the research instrument, one each on a separate card. Encourage the use of statements as opposed to actual questions to which the students can later agree or disagree with along a continuum scale. Encourage application of "the seven characteristics of a clear question" covered earlier in this chapter. Collect and shuffle the cards and begin to read each anonymous statement aloud for assessment by the entire classroom. Invite the students to apply the seven characteristics and make improvements to each question before proceeding to the next one.

Because the students are designing an instrument, encourage them to apply an additional criterion to the assessment, namely that of moderation. Remember that the best statements will be those that elicit a wide range of responses from strongly agree to strongly disagree. A statement to which everyone agrees such as, "The Bible is a helpful book for practical living" will add nothing to the variance of scores and thus add nothing to the comparison of scores. Shifting the question slightly toward, "Every religion in the world would benefit from reading and applying biblical principles to their lives" might add a little more variance to the range of responses.

If time allows, create a research instrument from the refined statements and distribute the instrument to the class for response. Include a few demographic questions for comparison. Collect the completed questionnaires and analyze them according to the practices suggested in the next chapter.

ADDITIONAL THOUGHTS AND RESOURCES

Remember that the research design should be firmly guiding the methodology at this point of the research study. The student's passion should inform the selection of the topic, which in turn should influence the choice of context, which should help the student surface the theological basis, which should shape the research question, which should dictate a preferred methodology, which in turn should drive the selection of the research procedures. Notice not only how each decision enlightens each subsequent decision, but also how the degree of influence grows the further that the student progresses through the funnel. At this stage of the research study, the choice of procedures should be fully defined by the previous selections of topic, context, and question. It is imperative at this point to select the most appropriate research procedures called for by the research design and the research question.

Each research procedure has a particular desired outcome. The research question should contain the starting point for your study and at least a hint at your desired destination. The research procedures are the vehicles that will transport the researcher between these two points. Allow your previous decisions to select the most appropriate procedures for your research design based upon the purposes of each procedure as outlined below and then conduct the procedure in such a manner that the purpose of the procedure is achieved.

1. Interview—the purpose of an interview is to gain new knowledge from a person who appears to be in the best position to reveal that information. Look for new insights and surprises during the interview. Pay attention to emotions and feelings. Watch for contradictions among interviewees that raise questions about previous learnings.

2. Focus Group—the purpose of a focus group is to gather the group's perspective on the topic. The best focus groups have a commonality and include participants who are anonymous to one another. Search for the unique perspective of this particular group. Limit questions to two or three per hour and allow the attendees to bounce ideas off one another until the group reaches a decision or consensus about a particular aspect of the topic before moving on to the next topic.

3. Observation—the purpose of conducting an observation is to gather individual pieces of information that later may be grouped into themes or used to confirm or contradict an emerging pattern from other data sources. Observe everything. Objects may be larger and carry more meaning than they originally appear.

4. Instrument—the purpose of an instrument is to measure a person's attitudes or beliefs about a particular topic. Try to find an existing validated instrument or carefully design your own. Use specific measurement protocols to avoid responder bias.

5. Survey—the purpose of a survey is to determine what is most substantive from the variety of information that you already have. Ask it all up front. Too many times, researchers collect data only to say at a later time, "I wish I would have asked . . ."

This chapter has lifted up a variety of best practices for employing research procedures. With data in hand, we now turn our attention toward best practices associated with data presentation.

Chapter 9

Presenting the Findings

THE CONCEPT

THE OTHER DAY I went to the movie theater and sat in my first push button reclining leather lounge chair complete with a matching red button to order beverages and food delivered to my personal location. I don't remember what I saw, but I do remember the lounge chair. I have always been enthralled with movie experiences. I still remember my first IMAX movie encounter thirty-five years ago. The camera placed the viewer in the heart of the action: in the front seat of a careening roller coaster, on the edge of a bouncing speedboat, and on the wing of a barnstorming biplane. Even if the viewer had never been in a biplane, a speedboat, or a roller coaster, the movie allowed viewers to feel as if they had experienced those events. That is what great movies do. That is also the goal of great data presentation.

The researcher should present the data in such a way that the reader feels as if he or she has lived through the same data gathering experiences as the researcher.

Detailing the Respondents

In order for the reader to discern the applicability of the findings, the reader must have a clear picture of who supplied the data. For qualitative data, paint a portrait of each respondent or group that was interviewed. Describe the backgrounds of the people. Detail the demographics. Include relevant idiosyncracies so that the reader receives a clear picture of the respondents. Research in a religious setting seldom involves comparing an experimental group with a control group. Since there is only an experimental group, that experimental group must be sufficiently defined in the eyes of the reader. As Miles and Huberman write, "Qualitative research is usually focused on the words and actions of people that occur in a specific context . . . Most qualitative researchers believe that a person's behavior has to be understood in context, and that context cannot be ignored or 'held constant.'"[1]

Provide as thick of a description of the respondents as possible without sacrificing their anonymity. Omit the actual names of the respondents and the organizations, or alter the names for referencing purposes in the text. Anonymity and context description are not mutually exclusive. Previously, in the research design phase of your research, you established the criteria for the selection of the participants. Now, tell us about the persons who were ultimately selected to supply the qualitative data for your project.

While the focus of the context centers upon groups and individuals in qualitative data, for quantitative data, the focus of the description is upon the larger organization. Much of the landscape may have already been painted by the description of your context earlier in the process. Details of the organization that were included in the contextual analysis do not need to be repeated here, but it is important to describe the specific setting in which the data was gathered. Answering the questions below may help the reader to gain a clearer picture of your setting.

1. Were the respondents already gathered for another purpose?

1. Miles & Humberman, *Qualitative Data Analysis*, 91.

2. How was the survey or instrument distributed?

3. How many people responded?

4. What measures were taken to ensure a higher, unbiased response rate?

5. If the data was collected in a single setting, were there any unusual circumstances on the day of the collection?

6. Where did the collection date fall in the church calendar year?

7. How did the attendance at the event compare to a typical day?

8. Were there any personnel changes recently?

9. What announcement did you give prior to distributing the questionnaire?

As much as possible, help the reader feel as if she or he was there for the data collection.

DIGGING IN

Presenting Qualitative Data

Themes

Remember the post-it note exercise described in the classroom exercise section of chapter 1? It may be helpful to refer back to that section of the book as you begin to generate themes from your qualitative data. The process used for constructing a definition of research from the individual and small group input of classroom participants is very similar to the exercise of constructing themes from qualitative data. In order to cluster specific pieces of data together, each piece of data must first be transferred to a card. Transfer every thought or concept that emerged from an interview, focus group, or observation to a 3 x 5 or 4 x 6 card. Use a new card for each new idea, but be sure to record a code on each card, such as a number corresponding to each respondent, so that you will be able to distinguish which ideas came from which respondents. The names of the respondents will not be revealed, but patterns will emerge from respondents who have one or more characteristics in common. If data was drawn from multiple groups, consider using different colored cards to represent the different commonalities for each grouping.

Once all of the pieces of data have been transferred to individual cards, begin to sort the cards according to the "alike of different" exercise explained in chapter 1. As we will discover, "Most words are meaningless unless you look backward or forward to *other* words."[2] The goal of this exercise is to group similar concepts together in order to generate a series of themes from the data. Once several thematic piles have been created from the data, begin to label each pile, giving the theme a name. Remember to review solo cards for possible inclusion into one of the named piles. When presenting the themes, list the name of the grouping first and then list the type of data that has been grouped into that theme. It is not necessary to list every card for every theme, but rather give the reader a sense of what each theme entails. The individual cards should support your newly given titles for the various groupings.

I recommend repeating this same exercise a couple of times in an effort to create new themes from the data each time. For instance, the first set of themes that emerge may be sequential in nature, such as the ones that typically emerge from the definition of research process itemized in chapter 1. Repeat the process a second time looking more for conceptual themes and perhaps a third time looking for space or cultural concepts. At this point in time, ignore patterns that may be emerging from the data according to the characteristics of the respondents themselves; that process will come next. Just pay attention to the data on the cards and ignore who said what. Record every theme that emerges from the data, but do not feel the need to present every emerged theme, lest the researcher portray a false sense of the amount of data collected.

Rest with your themes and then return to them a few times seeking to surface the themes that you believe portray the best picture of the qualitative data that you have collected. Once you

2. Ibid., 54.

have chosen the categories that you believe best portray the data that you have collected, gather feedback on the groupings. The best audience to supply feedback consists of the people who supplied the data in the first place. This may or may not be possible given your circumstances. Other options for gathering feedback include a group of your peers, a group of experts, etc. Consider asking the following questions for feedback:

1. What can you confirm?

2. What surprises you?

3. What is missing?

For a more formal categorization of themes, refer to Neuman,[3] who suggests conducting three subsequent analyses of the raw qualitative data, each adding an increasing level of detail to the categorization. He terms the first pass-through, "open coding," which involves, "looking for critical terms, key events, or themes."[4] He terms the second level analysis, "axial coding" which, "moves toward organizing ideas or themes and identifies the axis of key concepts in analysis."[5] The third pass-through, "selective coding," involves a re-coding of the raw data into the themes and sub-divisions identified in the second round.

Patterns

Themes refer to the categories created from the data itself. Patterns refer to trends that emerge from the data by analyzing who said what. Once you have chosen and labelled the themes of your data, begin to examine who said what within each theme. If you have incorporated cards of varying colors to represent your groupings, some of the patterns may jump out at you. For instance, do the older respondents tend to congregate into one pile and the younger respondents into another? Did the responses from the

3. Neuman. *Basics of Social Research*, 329–32.

4. Ibid., 330.

5. Ibid., 331.

males cluster into a few themes and the females into another set? Harken back to your research design and your literature review. If the literature suggested certain potential demographic differences in your data such as age, gender, tenure, location, ethnicity, etc., look for these patterns in the data that you collected. Do the anticipated patterns hold true? Do they not hold true? Report the findings whether or not the discovered patterns coincide with the anticipated results.

Relationships across Categories

As themes and patterns emerge from the data, consider a visual representation of the named themes and patterns. Create new 3 x 5 cards of the named themes and patterns and conduct a new alike or different process on the category titles themselves. Does the overall data suggest a process? Does a flowchart appear to be arising from the data? Does the data appear to be cyclical in nature? Is there one theme from which all others seem to emerge? This may also be an opportune time to return to the literature review to compare your findings with other perspectives, views, and theories. It could be that a particular theory that originally seemed disconnected now appears to be related to your study as a result of the specific themes that have now been identified.

Presenting Quantitative Data

Tables and Charts

Qualitative data is presented in the form of lists, diagrams, themes, and patterns. Quantitative data is presented in the form of tables and charts. When presenting quantitative data, the reader can often gain a more accurate picture of the data when presented closer to its raw form as in actual frequency counts and visual representations of those frequencies. A frequency count is simply a number indicating how many persons responded to a particular question in a certain way. A variety of tools are available to researchers to display their quantitative data to their constituents. Let's construct a sample data set and display the results in table form.

Suppose that the class has designed their own survey instrument related to the sense of community that a congregant may feel as a participant in a congregation. Presume that the instrument contains twenty-four questions or statements, each of which is to be answered on a one to five Likert scale according to the ratings of strongly agree, agree, neither agree nor disagree, disagree, and strongly disagree, producing a possible score ranging from twenty-four to one hundred and twenty for each person completing the instrument. Consider further that the designed instrument contains eight questions each in three dimensions of sense of community, namely supporting one another, willingness to spend time together, and respect for one another. The table below shows the scores, in the form of cell means, of a fictitious congregation in which three different age cohort groups are compared on the overall score of sense of community as well as its three dimensions.

Sense of Community Dimension	Length of Tenure		
	Less than 5 years n=10	5 to 10 years n=10	More than 10 years n=10
Support	28.1	32.4	37.0
Time	22.6	27.0	34.8
Respect	32.2	33.3	36.1
Total Score	82.9	92.7	107.9

Notice how the content dimensions are contained in the rows of the table and the comparison groups are depicted through the columns. Whenever constructing a table, the content should be shown in rows and the people in columns. This is easily remembered by thinking of each column as representing a person standing upright. Also notice how the numbers are centered in each column for easier reading. Using the same imagery, consider that the person is standing in the middle of the column rather than leaning to one side. The "n = 10" expression in each column indicates that a total of ten persons in each age group responded to the instrument. Use a small n when referring to subgroups and a capital N when referring to the total number of respondents, which in this case would be depicted by "N=30." It can also be helpful to shade the headings of the table or shade every other row if the table contains several rows.

Another way to show raw data that is sometimes used in instruments is to show the number of persons in each group who responded with a high or low rating. This same table could be reconstructed showing the number of persons who responded with a four or a five within each dimension of the survey. Remember that frequency counts are simply a measure of how many people responded in a certain way. It is fine to use a subset of the data so long as this portrayal passes the integrity test. A portrayal of data passes the integrity test if the intent of the portrayal is to display a more accurate and understandable representation of the data rather than a distorted view of the data.

Charting Data

Once the data is depicted in table form, many software programs will automatically convert the data in the table to a visual chart. One of the more popular forms of charting is called a histogram. "A histogram is nothing more than a pictorial representation of the frequency table."[6] A bar chart is similar to a histogram, but with some space in between each column. There are other forms of charts as well. A frequency polygon connects the midpoints of the heights of each column with a line segment. A pie chart shows the relative frequencies of the data by dividing a pie into segments representing the frequencies.

Some may have heard that, "you can make data say anything in tables and graphs." While that is true, a few rules will keep the researcher honest and achieve the goal of helping the reader understand the data as if he or she were there when it was gathered.

6. Kuzma, *Basic Statistics*, 31.

When displaying data in a bar chart or histogram format, always start both the horizontal and vertical axes at zero. A general rule of thumb is to make the height of a bar chart or histogram equal to three fourths of the length of the chart. Doing this will allow the area underneath the curve or bar of each of the columns to accurately represent the percentage of responses attributed to that column, providing a more accurate portrayal of the data. Remember how we pictured a person standing in each column? Increasingly larger columns represent an increasing number of people who responded to the chosen questions in a particular manner.

Other Data Measures

While frequency counts often provide the best window into the data itself, it is also appropriate to report the central tendencies as well as the ranges of the data set. Central tendencies are calculated or identified averages of the data. A mean is calculated by dividing a total score by the number of participants in the group. The median of a group of numbers is simply the middle number when the numbers are arranged in ascending or descending order. The mode of a group of numbers is the number that occurs most frequently. All three of these are measures of central tendency. It is also helpful to include the range of scores for each group, which merely translates to the highest and lowest score for each group.

Triangulating Data

When people get triangulated, that is usually not a good thing, but triangulating data is entirely appropriate and even desired. "Detectives, car mechanics, and general practitioners all engage successfully in establishing and corroborating findings with little elaborate instrumentation. They often use a *modus operandi* approach, which consists largely of triangulating independent indices . . . this is precisely how you get to the finding in the first place—by seeing or hearing multiple instances of it from different sources, and by squaring the finding with others it needs to be squared with."[7] As themes and patterns begin to emerge from the data, some of these themes and patterns occur multiple times. At this stage of the research, the researcher should merely note this in the findings. The significance of multiple occurrences will be covered in the next chapter. When noting evidence of triangulation or the proliferation of data, it is appropriate to look for signs of convergence both within varying sources of qualitative data such as observations, focus groups, and interviews as well as across qualitative and quantitative procedures.

CLASSROOM EXERCISES

Statistical Significance

Recall that in the classroom exercise, I suggested that each student be asked to rate her or his comfort level with statistics on a scale of one to ten, where a one represents extreme anxiety and a ten represents extreme comfort. Imagine that ten numbers in the next row represent the comfort level with statistics responses of a classroom of ten students.

7, 4, 8, 5, 5, 6, 7, 9, 4, 6

Would the reader say that the set of numbers just given differs from the next set of numbers below?

7, 4, 8, 5, 5, 5, 7, 9, 3, 5

In previous classroom settings the general consensus that typically surfaces regarding two sets of numbers like these is that while the two groups are not identical, the two groups are not substantively different from one another. As a classroom exercise, I continue to post new sets of ten numbers on the board, ensuring that each set is a little further apart from the original set of numbers, each time asking, "Is this new set of numbers substantively different from the original set?" Rather than arriving at a relaxed group consensus, most classes get frustrated over this exercise, suggesting that there should be a more definitive way to conclude

7. Miles and Huberman, *Qualitative Data Analysis*, 234.

that two groups are different from one another than just eyeballing the figures. There is.

There are a variety of statistical procedures that may be performed on two sets of numbers to determine whether or not two groups are substantively or significantly different from one another. While the procedures require certain assumptions or characteristics of the data to be satisfied in order for the calculated conclusions to be valid, the conclusions are based on similar theoretical bases.

Imagine that we randomly generated a set of ten numbers that range from one to ten one hundred different times, so that we have one hundred sets of numbers similar to the original set of ten numbers, except that each of these new sets was generated randomly. Imagine further that we compared the average distance of each set with every other set and that we are looking for a number large enough that only five of the one hundred groups differ from the other groups by that amount or more. Theoretically we could come up with a number that would suggest that if two groups of ten persons differ by this amount, we could conclude that there is a significant difference between these two groups because only five times out of a hundred will two groups differ by this amount when constructed randomly.

These imaginings are not merely hypothetical. It is possible to perform a statistical calculation to determine whether two groups are significantly different from one another. When we conclude that the two groups differ from one another based upon statistical analysis, we conclude that we are 95 percent sure that the difference between the two groups is not due to chance or randomness. There will still be a 5 percent chance that the difference between the two groups is due to random error, but most would agree that performing this calculation is preferable to eyeballing the data for differences. Threshold levels of greater than and less than 5 percent may also be used, but 5 percent is the standard threshold for statistical tests of significance.

Comparing Groups

Why would we want to show that two groups differ from one another? Suppose that the sense of community instrument discussed earlier was distributed to the same group of people at the beginning and end of a project and that in between the two collection times that several interventions were conducted to try to enhance the sense of community in that congregation or organization. Conducting a statistical procedure to determine whether the pre group differs from the post group would greatly enhance the conclusions that could be drawn from this study. Or imagine that the literature review suggested that males and females typically differ from one another regarding sense of community. This hypothesis could be tested using a group comparison analysis.

The most common group comparison analysis is the t-test. By computing a t-test, the researcher can determine whether two independent groups, such as males and females or members and non-members differ from one another. For multiple categories, such as age cohort groups, or length of tenure groups, an ANOVA is a very common analysis to run to compare whether multiple groups differ from one another. The most important piece of information from performing measures of comparison is the alpha level. An alpha level of less than 5 percent indicates that there is less than a 5 percent chance that the difference between the two groups is due to chance, and thus that they are substantively different from one another. An alpha level greater than 5 percent indicates that we cannot conclude that there is a difference between the two groups.

Nonparametric Tests

I mentioned earlier that certain characteristics must be satisfied in order to perform certain statistical analyses. A t-test is known as a parametric test. Parametric tests are based upon the assumption that some of the features of the sample group are similar to other groups in the overall population. One of these features is that the data is distributed in the form of a bell curve with most of the scores clustering in the middle and very few scores clustering at the extremes. Another assumed feature is that the mean, median, and mode of the data are all the same. If these features do not hold true, the results of the tests may be inaccurate. In very small data sets, it can be difficult to demonstrate that these features are true and thus more difficult to demonstrate statistical significance.

Fortunately, there are other procedures that can be run that are not built upon this same set of assumptions. This other set of procedures is known as nonparametric procedures. Nonparametric calculations do not require that this same set of assumptions be satisfied because they are based on ordinal rather than interval data. See chapter 7 for a discussion of ordinal and interval data. Nonparametric alternative analyses to the t-test include the Wald-Wolfowitz runs test, the Mann-Whitney U test, and the Kolmogorov-Smirnov two sample test. The Kruskal-Wallis analysis of ranks is an alternative for comparing multiple groups. For comparing related groups such as the pre and post groups in the sense of community study, Wilcoxin's matched pairs test is an alternative to the parametric t-test for dependent samples. Making pre and post comparisons is a special form of related samples testing in which each person serves as their own control for the two groups.

Because so many fields employ statisticians it is ofen relatively easy to find someone who will calculate one of the above analyses for a small fee. When approaching such a person, it is important to have the entire data set in spreadsheet form and to know the precise analysis desired and why. The results may include a variety of pieces of information, but the most important piece of information is whether or not the test was significant as indicated by the alpha level of the test. The researcher must still be able to interpret the results of a statistical analysis even if performed by a third party.

ADDITIONAL THOUGHTS AND RESOURCES

Measures of Variation

Let's return to the comfort level with statistics responses again. In what ways do the two groups of ten responses differ below?

5, 5, 5, 5, 5, 5, 5, 5, 5, 5

1, 1, 1, 1, 5, 5, 9, 9, 9, 9

While the two groups have the same total of fifty and same mean of 5.0, they differ in another way quite dramatically. The range of the first group is zero, while the range of the second group is eight. The second group of numbers is much more spread out than the first group. In addition to presenting figures of central tendency such as the mean or median, another concept that is helpful to show to the readers are measures of variation. The most common measure of variation is the standard deviation. It is computed by squaring each deviation from the mean, adding them up, and dividing the sum by one less than n. Don't worry about memorizing this calculation. Most software programs will calculate a standard deviation for you once the data is in spreadsheet or tabular form.

While it is not necessary to know how to compute a standard deviation at this level, it is important to be able to interpret it. The larger the standard deviation, the greater variability will be present within the data and the smaller the standard deviation, the less variability will be present within the data. Calculating a standard deviation for a sample size rather than an entire population requires a different computation, producing what is known as a standard error rather than a standard deviation.

Measures of variation are completely separate concepts from measures of central tendency, such as mean, median, and mode. Data reporting should always include both of these concepts. Another favorite classroom exercise of mine is to require students to explain the differences between measures of central tendency and measures of variation in data to someone who knows very little about statistics. Over the years, some very interesting conversations have been reported from this assignment as students attempt to explain these concepts to Aunt Mable or Uncle John.

Type I and Type II Errors

Earlier I mentioned that there is a small chance that a statistically significant conclusion could be wrong. There are actually two types of errors that can be made. The first error can occur if we conclude that something is false when it is actually true. The second error that can be made is concluding that something is true when it is actually false. I once delivered a sermon on type I and type II errors. Pity the poor congregation that endured that sermon! Actually, examples of these elements are quite prevalent in Scripture. As another classroom exercise, I have asked my students to find biblical examples of type I and type II errors. It is surprising how many examples of type I and type II errors are evidenced in Scripture. For all of the analyses that can be conducted, drawing conclusions is still a matter of discernment rather than a matter of calculation.

Chapter 10

Significance of the Study

THE CONCEPT

WE HAVE NOW REACHED the bottom of the funnel. Up until this point in the funnel, the funnel has become narrower with each subsequent step in the study. Hopefully the reader has discovered how the passion, context, biases, and dreams of the researcher must drive the methodology of the study and never vice versa. The passions of the researcher must drive the procedures conducted, lest the study be tainted or constrained needlessly by irrelevant or untimely concerns.

As water flows through a funnel, the water gains speed as it pours through an ever smaller opening and experiences the pressure of the water behind it. If there is too much pressure, the water can experience turbulence, actually slowing the rate of speed. You may be familiar with a study that experienced turbulence in the middle. But, most research studies gain momentum as progress is made in the study. Reaching the point of data collection can be an exhilarating midpoint giving new energy to the entire system. This momentum, however, seems to reach a grinding halt as one pursues the significance of the study. There is good reason for this. The pace slows because for the first time since the early stages of the study, the landscape has broadened rather than narrowed. Naming significance is concerned with how the study will influence not only this organization but potentially many other organizations. Water is no longer flowing just through your organization or the set of organizations in your study. We now need to release the water to flow out to others. Such work will require patience, wisdom, and reflection.

There is a story told of Isaac, son of Aaron who lived in the Polish city of Krakow.[1] Isaac was normally a very sound sleeper. He worked a laborer's job, was often very tired at night, and slept well. But lately, Isaac had been dreaming a recurring dream. He dreamed the same dream every night for two weeks. In his dream, he dreamed of a treasure hidden under a bridge in the far off city of Prague. Exhausted from lack of sleep, Isaac decided to traverse the arduous journey to Prague to satisfy his curiosities so that he could return to sleeping soundly. At the end of his three day journey, he came upon the bridge in his dreams. Every detail was just as he had dreamed it to be. As he was about to descend into the river, a stern police officer stopped and questioned him. Because Jews were seldom seen in this part of the city, the interrogation continued at the police station. With no other defense than the truth, Isaac revealed his recurring dream. "You are silly to believe in dreams," declared the officer, "Why, for the last two weeks I myself have dreamed that in the far off city of Krakow, in the home of a peasant, Isaac, son of Aaron, there is a treasure hidden under a stove in the kitchen, but you don't see me wasting my time following dreams."[2] The officer would not allow Isaac to return to the bridge over the Vltava River. He simply walked back to his home in Krakow, moved the stove in the kitchen, and found a buried treasure.

Sometimes treasures are hidden under our noses, but only a journey can make them visible to us. Research endeavors are journeys. They take us to new places, introduce us to new people, and provide us with new information. Much of the neoteric information can be significant, but so, can the information that was

1. White, *Stories for Telling*, 51.
2. Ibid., 52.

very close to us all along. The last chapter in this journey is about discovering the significance of one's study.

Significance can be found in new ideas and new theories, as well as in seeing the familiar in a new light.

Remember that asking questions is like planting seeds. We can never fully anticipate how, where, or even why a seed will grow. Well-designed research studies surface both new information and new perspectives. Sometimes a respondent reveals a piece of information that even the person giving the response was unaware of until it surfaced. Good questions penetrate the soil, good analysis harvests what grows, and good reflection reveals what is significant. Reflection is the key to surfacing the significance of a study, but reflection requires time. Pulling on a plant is not a healthy way to increase its growth. Thankfully, for the researcher of religious studies, reflection is a two way street.

In Matthew 13:44 we read, "The kingdom of heaven is like treasure hidden in a field, which someone found and hid; then in his joy he goes and sells all that he has and buys that field. Again, the kingdom of heaven is like a merchant in search of fine pearls; on finding one pearl of great value, he went and sold all that he had and bought it." Jack Sanford points out that these two parables are not parallel. "At first glance," he writes, "it looks as if this parable duplicates the parable of the treasure. But, as Fritz Kunkel has pointed out, in the first parable the kingdom is a treasure that we search for and find; in the second parable the kingdom is likened to a merchant who is searching for things of value. In this case we are the pearls, the treasure found by the kingdom of God."[3] The good news is that God is searching just as actively to reveal the significance of the study to us as we are actively searching for it.

---

3. Sanford, *Kingdom Within*, 27.

DIGGING IN

Steps to Surfacing Significance

Searching for buried treasure is a journey, not a task. Isaac dreamed a recurring dream for two weeks until he finally acted upon it and even then the journey was just beginning. Searching for the significance of the study is one part of the study that cannot be rushed. Like all other processes in this book, however, the process can be framed. There are steps that can be taken to discover the significance of the study. Begin by reviewing every piece of the overall findings. What do the tables reveal? What do the themes disclose? Name the obvious. Were your assumptions and expectations confirmed? If not, why not? Next review the triangulation of your study. Do the various sources of information point in the same direction? Are there subtle differences in where the vectors lead you? What is responsible for those differences? Thirdly, review the outliers, those pieces of data that simply do not fit anywhere else. Did someone offer a perspective offered by no one else? Did one particular focus group take you into new territory? Did someone's journaling offer deeper insights than what the other data revealed? Return to those pieces of information that are still unique in their relationship to the overall gathered data.

Sources of Significance

Significance tends to surface from three main areas, namely the main points, the minor points, and the pathways to the points. Main points that seem obvious can cry out for significance because they are astonishing or insightful as in the case of the former student's study that revealed why some crisis workers are able to endure for many years while others do not get past the first year or two. One of the key findings of this study was that lack of ambiguity in one's faith was a strong determinant to allowing a person to persevere in this field. Findings can turn into significant outcomes due to the nature or magnitude of the finding. In this case, the finding was significant because it showed that faith matters in a role held by many secular persons. Recall the sense of community study that we have used as a fictitious case throughout this book. Suppose that the study contained five age groupings and that every age group revealed that "respect" was the key to building a sense of community. In that case, the finding would carry significance due to its magnitude among the age groupings. Main point findings also can be significant because they are unexpected. In an earlier study of mine, conducted when I was serving as pastor of the congregation, I discovered that the oldest age cohort had the least ties to tradition in the congregation. That finding proved to be significant in advancing the mission of the congregation.

Significance can also surface from the minor points rather than the main points of the study. One former student of mine was attempting to show a relationship between pastoral emotional strength and the health of congregations. The emphasis of the study was placed upon the linkages between the pastor and congregation and several interventions were made to increase the emotional strength of the pastors involved in the study. The researcher assumed that the interventions would increase the emotional strength of the pastors, but was much more uncertain about the link between emotional strength and congregational health. The results of the study revealed something unexpected. The emotional strength of a sizable portion of the participants actually decreased rather than increased during the study! I still remember his first question to me following the data discovery, "Do I throw this out and start over?"

Churches are not petri dishes. While the result could have been the result of a type II error, there was probably a good reason for the finding. He needed to dig deeper into the findings. Thankfully he had asked the participants to keep a journal during the study. Through interviews conducted of the participants with journals in hand he was able to determine that several of the participants had raised the bar of their understanding of emotional strength during the study. While many believed that they had grown in their emotional strength, they also believed that they had further to go to reach maturity in this area than when they had initially commenced the study. One of the conclusions of his study suggested that understanding emotional strength is as important as seeking to advance it among pastors.

A third source of significance can be found in the pathways to the findings. Sometimes the findings of a study appear obvious, but the pathway to them reveals significance in the study. A colleague of mine recently completed a study on women in ministry in which the barriers to accessing positions of ministry were revealed. At first glance the themes that emerged to overcoming barriers to access seemed expected, two of which were: "know your call" and "network." Further digging into the qualitative data revealed that women often go about these tasks differently than men. While most women in the study were able to claim their calling, many of the women who served as respondents in this study had not rehearsed their call story due to the long and complicated nature of the call. Not rehearsing one's call can send signals of doubt to a potential search committee. Further digging into the networking access point revealed that the women in this study often did not act upon the notion of networking with their contacts even though they acknowledged that networking was a significant pathway to accessing a call. They considered it presumptuous to ask a friend of theirs about an open position. A third access point for women in ministry, that of simply applying for a position, also revealed significance in the pathway to this finding in that women often do not apply for a position unless they can demonstrate that they meet all of the requirements, a finding that is not as true for their male colleagues.

Another source of possible significance in studies can surface from the potential impacts raised by the study. Cahalan distinguishes between three types of impact as outlined below.[4]

1. Initial—the immediate benefit that comes from participation.

---

4. Cahalan, *Projects That Matter*, 18.

2. Intermediate—changes in knowledge, attitudes, skills, and behaviors that establish new patterns of thinking and behaving.

3. Long-term—Changes in identity, a condition or status that persists over time.

Sometimes, studies surface changes in people as outlined in Cahalan's categories. Such changes can be significant. Even greater are new identities that surface in people or organizations. In defining the significance of the study, Cahalan's typology may be used to categorize the potential impacts of the study and suggest conclusions. Drawing conclusions is an important part of surfacing the significance of the study. In many cases, conclusions simply come in the form of insights that have been gleaned from the main points, minor points, and pathways of the study. For other studies, as in the case of a recent student of mine who studied various preparatory tracks to ministry for African American pastors, conclusions need to be presented in the form of recommendations.

CLASSROOM EXERCISE

An exercise that can be performed in a classroom setting is to pose questions of a particular researcher who has recently completed his or her study. This can be a great learning opportunity for those just forming their own research designs. Below, I have included a set of questions that can be posed to a researcher who is rejoicing over just having completed a study.

1. What was your original intent for the study? Has it been met?

2. What have been the unintended consequences or side effects of the study?

3. What was the overall philosophy behind the study? Did the philosophy help or hinder the study? Was the philosophy consistent with those who participated?

4. What were your biases in the study? How did they affect the study?

5. How did the study advance the mission of your organization?

6. How might you distinguish between the merit (overall value) of the study and its worth (specific to your organization)?

7. What were the tradeoffs for being involved in this study for you personally? (How else would the same energy, time, and resources be utilized if not directed toward this study?)

8. What about the trade-offs for the organization?

9. Who benefited most from the study? Its leaders, its followers, or the organizations themselves?

10. Does the organization need to gain ownership over some aspect of this study going forward? How might that happen?

11. How did you change personally during this study?

12. What advice would you have for someone wanting to conduct a similar study?

13. What is your next area of study?

14. How might the curriculum of this educational institution be improved?

ADDITIONAL THOUGHTS AND RESOURCES

Further Study

The final sections of a study should also include suggestions for others who wish to build upon your study. Should your study be replicated in a different type of setting? Did you uncover some minor points or pathways that should become the primary focus of a future study? Should the next endeavor embrace a broader, more qualitative approach or a more in-depth, quantitative approach?

Cross-Cultural Settings

The funnel approach to research design, upon which this study is based, has been used in multicultural as well as cross-denominational settings. I have taught this same approach to students at the Myanmar Institute of Theology of Burma, and with some cultural adjustments, the process worked quite well. I believe that the adaptability of this approach lies within its core that is based upon concepts rather than language or formulas. The approach has also been used in a Korean setting and with institutions that are primarily Baptist, Lutheran and Methodist. Students of mine in these institutions have included every mainline tradition as well as many other Christian traditions. I welcome feedback on the potential adaptability or lack thereof in other settings.

Epilogue

I leave the reader with ten insights that you may not receive in a traditional classroom setting.

1. Your advisor has final say in your research study. It is never a good idea to try to trump what your advisor says even if the information comes from a reputable source such as from your research design or research methods professor.

2. Your study should seek to make a difference but stop short of attempting to change the world.

3. Passion matters but so does the passion of your faculty and advisors.

4. It is wonderful to identify your topic early in your curriculum, but try to get your study approved prior to collecting any data.

5. If you are not involved in designing research for a PhD program, your study should not resemble one.

6. Every new phase of your project should cite new bibliographic material.

7. Completing a degree gives your same words more weight after the degree is completed.

8. If done right, your research study will launch rather than end your research career.

9. Your spouse, friends, and family are needed for support. But they are not your research committee.

10. Treat your advisor as your mentor. You are still learning to do research. When you complete your study, you can conduct the next one with no one looking over your shoulder.

11. It is rare to publish your research study in its final form, but you should try to publish something from the study.

12. Expect three significant things to go wrong with your study and do not panic when they do.

Bibliography

Ammerman, Nancy, et al. *Studying Congregations*. Nashville: Abingdon, 1998.

Axelrod, Richard. *You Don't Have to Do It Alone: How to Involve Others to Get Things Done*. San Francisco: Berrett-Koehler, 2004.

Bevans, Stephen. *Models of Contextual Theology*. Maryknoll, NY: Orbis, 2011.

Branson, Mark. *Memories, Hopes, and Conversations: Appreciative Inquiry and Congregational Change*. Bethesda, MD: Alban, 2004.

Brown, Juanita and David Isaacs. *The World Café: Shaping our Futures Through Conversations that Matter*. San Francisco: Berrett-Koehler, 2005.

Cahalan, Kathleen A. *Projects That Matter: Successful Planning and Evaluation for Religious Organizations*. Bethesda, MD: Alban, 2003.

Chryssides, George, and Ron Geaves. *The Study of Religion: An Introduction to Key Ideas and Methods*. 2nd ed. London: Bloomsbery, 2014.

Creswell, John W., and and Vicki L. Clark. *Designing and Conducting Mixed Methods Research*. California: Sage, 2007.

Dannemiller, Kathie. *Whole-Scale Change: Unleashing the Magic in the Organization*. San Francisco: Berrett-Koehler, 2000.

Ernst, Chris, and Donna Chrobot-Mason. *Boundary Spanning Leadership: Six Practices for Solving Problems, Driving Innovation and Transforming Organizations*. New York: McGraw Hill, 2010.

Fowler, Floyd J. *Survey Research Methods*. London: Sage, 1988.

Guba, Egon G., and Yvonna S. Lincoln. *Naturalistic Inquiry*. Newbury Park, CA: Sage, 1985.

Harrison, Owen. *Open Space Technology: A User's Guide*. San Francisco: Berrett-Koehler, 2008.

Heifitz, Ronald, et al. *The Practice of Adaptive Leadership*. Cambridge, MA: Harvard University Press, 2009.

Jones, Laurie Beth. *The Path*. New York: Hyperion, 1996.

Jung, Carl G. *Memories, Dreams, and Reflections*. New York: Random House, 1965.

Kondrath, William M. *God's Tapestry: Understanding and Celebrating Differences*. Bethesda, MD: Alban, 2008.

Kuzma, Jan. *Basic Statistics for the Health Sciences*. Mountain View, CA: Mayfield, 1992.

Lencioni, Patrick. *Silos, Politics and Turf Wars*. San Francisco: Jossey-Bass, 2006.

Ludema, James. *The Appreciative Inquiry Summit: A Practitioner's Guide for Leading Large-Group Change*. San Francisco: Berret-Koehler, 2003.

Marquardt, Michael. *Action Learning: Solving Problems and Building Leaders in Real Time*. Mountain View, CA: Davies-Black, 2004.

Martin, Roger. *The Opposable Mind*. Boston: Harvard Business School Press, 2007.

Miles, Matthew B., and Michael A. Huberman. *Qualitative Data Analysis: A Sourcebook of New Methods*. Beverly Hills: Sage, 1984.

Moustakas, C. *Phenomenological Research*. London: Sage, 1994.

Neuman, W. Lawrence. *Basics of Social Research*. 2nd ed. Boston: Allyn & Bacon, 2004.

Pearson, Carol. *Awakening the Heroes Within: Twelve Archetypes to Help Us Find Ourselves and Transform Our World*. San Francisco: Harper, 1991.

————. *Organizational and Team Culture Manual*. Gainesville, FL: Center for Applications of Psychological Type, 2004.

Pearson, Carol, and Hugh K. Marr. *Pearson-Marr Archetype Indicator Instrument Manual*. Gainesville, FL: Center for Applications of Psychological Type, 2003.

Radwin, David. "High Response Rates Don't Ensure Survey Accuracy." *Chronicle of Higher Education*, October 9, 2009, B8–9.

Roozen, David, and James R. Nieman, eds. *Church, Identity, and Change*. Grand Rapids: Eerdmans, 2005.

Sanford, John A. *The Kingdom Within: The Inner Meaning of Jesus' Sayings*. San Francisco: Harper & Row, 1970.

Shute, Nancy. "The 18-Second Doctor." *U.S, News & World Report* 142/11 (2007) 14.

Snow, Luther K. *The Power of Asset Mapping*. Bethesda: Alban, 2004.

Sparks, Paul, Tim Soerens, and Dwight J. Friesen. *The New Parish: How Neighborhood Churches are Transforming Mission, Discipleship, and Community*. Downers Grove, IL: InterVarsity, 2014.

Swinton, John, and Harriett Mowat. *Practical Theology and Qualitative Research*. London: SCM, 2011.

Walliman, Nicholas. *Research Methods: The Basics*. London: Routledge, 2011.

Warner, Stephen, and Judith Wittner. *Gatherings in Diaspora, Religious Communities and the New Immigration*. Philadelphia: Temple University Press, 1998.

Weisbord, Marvin R., and Sandra Janoff. *Future Search: An Action Guide to Finding Common Ground in Organizations and Communities*. San Francisco: Berrett-Koehler, 1995.

Wenger, Etienne, et al. *Cultivating Communities of Practice: A Guide to Managing Knowledge*. Boston: Harvard Business School Press, 2002.

Wheatley, Margaret, and Deborah Frieze. *Walk Out Walk On: A Learning Journey into Communities Daring to Live the Future Now.* San Francisco: Berrett-Koehler, 2011.

White, William R. *Stories for Telling: A Treasury for Christian Storytellers.* Minneapolis: Augsburg, 1986.

Woods, C. Jeff. *We've Never Done It Like This Before.* Bethesda, MD: Alban, 1994.

———. *On the Move: Adding Strength, Speed, and Balance to your Congregation.* Atlanta: Chalice, 2009.

Woodward, J. R. *Creating a Missional Culture.* Downers Grove, IL: InterVarsity, 2012.